Using Statistics to Understand the Environment

Introductory courses to statistics are the bane of many students' early academic lives. Yet the concepts and techniques involved are crucial for understanding the systems under examination. This is particularly true for environmental investigations, which cover a broad range of disciplines: from social science to pollution studies and from environmental management to ecology. Since these systems are complex, statistical methods are important techniques in the interpretation of project results.

Using Statistics to Understand the Environment covers all the basic tests required for environmental practicals and projects, and points the way to the more advanced techniques needed in more complex research designs. Following an introduction to project design, the book covers methods to describe data, to examine differences between samples, and to identify relationships and associations between variables.

Featuring: worked examples covering a wide range of environmental topics, drawings and icons, chapter summaries, a glossary of statistical terms and a further reading section, this book focuses on the needs of the student rather than on the mathematics behind the tests and provides an invaluable student-friendly introduction to this core area of Environmental Studies.

Philip Wheater is a Principal Lecturer in the Department of Environmental and Geographical Sciences at the Manchester Metropolitan University. **Penny Cook** is a Research Fellow in the School of Health at Liverpool John Moores University and an Honorary Research Fellow in the Department of Environmental and Geographical Sciences at the Manchester Metropolitan University.

Routledge Introductions to Environment Series, edited by Rita Gardner and A.M. Mannion.

Routledge Introductions to Environment Series

Published and Forthcoming Titles

Titles under Series Editors:
Rita Gardner and Antoinette Mannion

Environmental Science texts

Environmental Biology
Environmental Chemistry and Physics
Environmental Geology
Environmental Engineering
Environmental Archaeology
Atmospheric Processes and Systems
Hydrological Systems
Oceanic Systems
Coastal Systems
Fluvial Systems
Soil Systems
Glacial Systems
Ecosystems
Landscape Systems
Natural Environment Change
Using Statistics to Understand the Environment

Titles under Series Editors:
David Pepper and Phil O'Keefe

Environment and Society texts

Environment and Economics
Environment and Politics
Environment and Law
Environment and Philosophy
Environment and Planning
Environment and Social Theory
Environment and Political Theory
Business and Environment

Key Environmental Topics texts

Biodiversity and Conservation
Environmental Hazards
Natural Environmental Change
Environmental Monitoring
Climatic Change
Land Use and Abuse
Water Resources
Pollution
Waste and the Environment
Energy Resources
Agriculture
Wetland Environments

Energy, Society and Environment
Environmental Sustainability
Gender and Environment
Environment and Society
Tourism and Environment
Environmental Management
Environmental Values
Representations of the Environment
Environment and Health
Environmental Movements
History of Environmental Ideas
Environment and Technology
Environment and the City
Case Studies for Environmental Studies

Routledge Introductions to Environment Series

Using Statistics to Understand the Environment

C. Philip Wheater
and
Penny A. Cook

Illustrations by Jo Wright

Routledge
Taylor & Francis Group

LONDON AND NEW YORK

First published 2000
by Routledge
2 Park Square, Milton Park, Abingdon, Oxon, OX14 4RN

Simultaneously published in the USA and Canada
by Routledge
270 Madison Ave, New York NY 10016

Reprinted 2002, 2003

Transferred to Digital Printing 2006

Routledge is an imprint of the Taylor & Francis Group

© 2000 C. Philip Wheater and Penny A. Cook; Jo Wright for the illustrations

The right of C. Philip Wheater and Penny A. Cook to be identified as the Authors of this Work has been asserted by them in accordance with the Copyright, Designs and Patents Act 1988

Typeset in Times by Keystroke, Jacaranda Lodge, Wolverhampton

British Library Cataloguing in Publication Data
A catalogue record for this book is available from the British Library

Library of Congress Cataloging in Publication Data
A catalog record for this book has been requested

ISBN 0–415–19887–9 (hbk)
ISBN 0–415–19888–7 (pbk)

Publisher's Note
The publisher has gone to great lengths to ensure the quality of this reprint but points out that some imperfections in the original may be apparent

Printed and bound by CPI Antony Rowe, Eastbourne

Contents

Series Editors' Preface
Environmental Science titles

The last few years have witnessed tremendous changes in the syllabi of environmentally related courses at Advanced Level and in tertiary education. Moreover, there have been major alterations in the way degree and diploma courses are organised in colleges and universities. Syllabus changes reflect the increasing interest in environmental issues, their significance in a political context and their increasing relevance in everyday life. Consequently, the 'environment' has become a focus not only in courses traditionally concerned with geography, environmental science and ecology but also in agriculture, economics, politics, law, sociology, chemistry, physics, biology and philosophy. Simultaneously, changes in course organisation have occurred in order to facilitate both generalisation and specialisation; increasing flexibility within and between institutions in encouraging diversification and especially the facilitation of teaching via modularisation. The latter involves the compartmentalisation of information, which is presented in short, concentrated courses that, on the one hand are self-contained but which, on the other hand, are related to prerequisite parallel and/or advanced modules.

These innovations in curricula and their organisation have caused teachers, academics and publishers to reappraise the style and content of published works. While many traditionally styled texts dealing with a well-defined discipline, e.g. physical geography or ecology, remain apposite there is a mounting demand for short, concise and specifically focused texts suitable for modular degree/diploma courses. In order to accommodate these needs Routledge has devised the Environment Series, which comprises Environmental Science and Environmental Studies. The former broadly encompasses subject matter which pertains to the nature and operation of the environment and the latter concerns the human dimension as a dominant force within, and a recipient of, environmental processes and change. Although this distinction is made, it is purely arbitrary and for practical rather than theoretical purposes; it does not deny the holistic nature of the environment and its all-pervading significance. Indeed, every effort has been made by authors to refer to such interrelationships and provide information to expedite further study.

This series is intended to fire the enthusiasm of students and their teachers/lecturers. Each text is well illustrated and numerous case studies are provided to underpin general theory. Further reading is also furnished to assist those who wish to reinforce and extend their studies. The authors, editors and publishers have made every effort to provide a series of exciting and innovative texts that will not only offer invaluable learning resources and supply a teaching manual but also act as a source of inspiration.

A. M. Mannion and Rita Gardner

1997

Series International Advisory Board

Australasia: Dr Curson and Dr Mitchell, Macquarie University

North America: Professor L. Lewis, Clark University; Professor L. Rubinoff, Trent University

Europe: Professor P. Glasbergen, University of Utrecht; Professor van Dam-Mieras, Open University, The Netherlands

Figures

Tables

Boxes

Worked examples

Acknowledgements

We would like to thank all those who fostered our early contact with statistics (especially Robin Baker, Gordon Blower, Lawrence Cook, Ian Harvey and Mike Hounsome). We are also grateful to the many colleagues and generations of students who have commented on earlier versions of this text and discussed aspects of our statistics teaching with us (in particular John Appleyard, Mark Bellis, Paul Chipman, Rod Cullen, Pete Dunleavy, Alan Fielding, Martin Jones and Mark Langan). Several reviewers commented on the initial proposal and a final draft of this text; we would especially like to thank Chris Barnard and the series editor Antoinette Mannion. A number of others have been kind enough to comment on drafts of this book and we would especially like to thank Roger Cook and Ian Harvey who persevered through the whole text, making copious helpful comments on the way. Despite their diligence, any mistakes remain our own. Professor Steve Dalton in the Department of Environmental and Geographical Sciences at the Manchester Metropolitan University kindly provided support during the writing of this book. We would like to thank all at Routledge involved with the production of this book, especially the copy editor, Richard Leigh.

The worked examples and questions in this book use data designed to demonstrate the methods in an appropriate way. Although the examples are inspired by real data from environmental situations, we caution the reader not to read any environmental conclusions into the analyses presented here. The data used have been selected to enable the technique to be demonstrated as clearly as possible, rather than to examine particular environmental situations. The following (together with various unpublished data of our own) were sources of many of these ideas: Bradshaw A. D. and Chadwick M. J. (1980) *The Restoration of Land*, Blackwell, Oxford; Brown A. (ed.) (1992) *The UK Environment*, HMSO, London; GEMS Monitoring and Assessment Research Centre (1991) *UNEP Environmental Data Report*, Blackwell, Oxford; Jenkinson S. and Wheater C. P. (1998) The influence of public access and sett visibility on badger sett persistence, *Journal of Zoology*, 246: 478–482; Prendergast S. and Wheater C. P. (1996) Access to the countryside by agreement: a study into the extent and nature of countryside recreational provision available through access to private land by agreement, Report to the Countryside Commission; Scott R. (1990) The conservation management of urban grassland, Unpublished M.Sc. thesis, Manchester Polytechnic;

Read H. J., Wheater C. P. and Martin M. H. (1987) Aspects of the ecology of Carabidae (Coleoptera) from woodlands polluted by heavy metals, *Environmental Pollution*, 48: 61–76; Wheater C. P. (1985) Size increase in the common toad *Bufo bufo* from Cheshire, *Herpetological Journal*, 1: 20–22.

The statistical tables are extracted from the following sources by kind permission of the publishers: Tables D.1–D.8 and D.12 are taken from Neave H. R. (1995) *Elementary Statistics Tables*, Routledge, London, and Neave H. R. (1995) *Statistics Tables*, Routledge, London; Table D.9 is taken from David H. A. (1952) Upper 5 and 1% points of the maximum *F*-ratio, *Biometrika*, 39, 422–424, Oxford University Press and by permission *Biometrika trustees*; Table D.10 is taken from Harter H. L. (1960) Tables of range and studentized range, *Annals of Mathematical Statistics*, 31, 1122–1147.

Every effort has been made to contact copyright holders for their permission to reprint material in this book. The publisher would be grateful to hear from any copyright holder who is not here acknowledged and will undertake to rectify any errors or omissions in future editions of this book.

Using this book

During environmental investigations we measure biological, chemical, physical and anthropogenic aspects of the internal and external environment. In many cases, these measurements form part of a monitoring programme describing the prevailing conditions: the pollutant levels in air and water; the number of species in a nature reserve; the density of traffic during peak periods. For such examples, a single absolute measurement may be sufficient to decide, for example, that current levels of emissions from cars or pollutants in the sea at a particular locality are within or exceed European guidelines. For other investigations, however, simply reporting a single measurement may not suffice.

Instead of finding one measurement to describe an aspect of the environment, we may wish to ask more sophisticated questions: to compare pollutant levels above and below an outflow of industrial effluent; to examine the effect of distance away from a pollution source on particulate lead levels; to establish which of several nature reserves has the richest flora; or to determine the time of day at which traffic flow is highest. We now enter the realms of data collection, investigation and interpretation; in other words, we need to design a project, collect data and do appropriate statistics. Only if these steps are carried out properly will we be able to answer our question.

Although statistical techniques are a necessary tool for many scientists and social scientists, not least for those working in environmental subjects, the ideas behind the methods and the heavy dependence on mathematics are often off-putting. However, no one would expect a successful car driver to have a degree in engineering or a computer operator to understand the development of silicon chips. Similarly, it is an understanding of the principles, rather than the details, of statistical methodology that is required for its successful use. Students must, however, understand when to use a particular method in order to avoid using inappropriate techniques which would be unable to adequately answer the questions posed. Worse, without some knowledge of statistical principles, the data may not even be collected in a way which allows proper interpretation.

This may begin to dismay those who still feel that the mathematical aspects of statistical techniques are beyond them. However, it is the intention of this book to cover the major groups of statistical tests in a way which allows users to select, employ and interpret the correct test for the correct occasion, whilst keeping the mathematical aspects at bay. This is especially relevant today, with the vast majority of users of statistical methods employing computer programs rather than calculation by hand.

The most effective way to learn how to use statistics as a tool within environmental investigations is to use the relevant techniques on appropriate data. Unfortunately, collection of data in a suitable way requires a knowledge of the statistical analyses to be subsequently used. This 'chicken and egg' situation means that, for most people, their first encounter with statistics is with data provided by someone else (hopefully collected using a sound methodology). No wonder statistics is all too often seen as a bolt-on extra to another course; a necessary chore without any real application. This book attempts to avoid such problems. It begins by introducing experimental design (Chapter 1), the principles of which should be understood even by those starting with class data or an experiment or survey designed by someone else. It is only with this understanding that researchers can critically decide whether the results generated by data really answer the intended question. As importantly, since statistics can be, and unfortunately often are, abused within the scientific, technical and social scientific literature, an appreciation of the way in which data are collected will enable the critical evaluation of the work of others.

The book continues with methods to summarise and describe data (Chapter 2), followed by descriptions of the commonly employed statistical tests (Chapters 3–7). As with most new subjects, an understanding of statistical methods involves getting to grips with some specialised terminology. Although we have tried to keep the jargon to a minimum, a number of terms and concepts are important. A glossary is included which cross-references the technical terms used to the relevant chapter (the first use of a glossary term in the text is highlighted in bold). For those who are a little more advanced, or who wish to delve a little deeper, formulae and worked examples for each test are included in the text. To help with this process there is a glossary of the notation used within the formulae and a brief explanation of the basic mathematical skills needed to compute tests by hand (Appendix A). In addition, all the statistical tables needed to complete the examples by hand are given in Appendix D. Although computers take away the drudgery of analysis, simple examples worked by hand offer insights into the logic behind the statistical test (note that small rounding errors may result in slight differences between the answers given in this book and calculations made by hand or using computer software). To assist the reader in tracking the worked examples through the book, appropriate drawings and icons flag those using the same data. The worked examples have been prepared to illustrate the workings of the methods as clearly as possible. However, this has sometimes meant that the data are presented in a layout which may not be suitable for analysis by computer. Since many of the users of this book will indeed analyse their data using computer programs, for each statistical test there are illustrations of the usual formats for entering data in the most commonly used statistics programs (Appendix B). Each section ends with questions which reinforce statistical skills. Answers to these problems are given at the end of the book (Appendix E).

The first-time user is advised to work through this book in a systematic way. For those with some knowledge, or for whom the book provides a refresher course, chapters (or sections) may be read in isolation. A quick guide to the range of tests covered by this book can be found at the very end of the book. Having carried out a successful project and analysed the data, the final hurdle is reporting the analysis in the results section of a report or paper. Because statistical output can seem complex, throughout the book, those elements which should be included in a results section have been highlighted. This includes suggested forms of words, together with examples of the styles of tables and figures which could be used to display the results.

This is an introduction to statistics. At each stage, when the limit of this book is reached, brief descriptions of more advanced tests are given, together with a source of reference. This, together with the further reading section (Appendix C), should allow the reader to be aware of a wider range of tests than can easily be covered in a text of this length, and point them in the direction of more detailed reference books. More experienced users may find that this book acts as a first port of call in deciding which test is the most appropriate for the question being addressed.

① Project design

Careful planning and implementation of data collection are fundamental to the success of any project. The links between the development of a project design and the subsequent analysis and interpretation of the work are dealt with in this chapter, especially in relation to:

- **Schemes of sampling and experimental layout**
- **Using questionnaires, and semi-structured and unstructured interviews**
- **Sources of error**
- **Types and properties of data**
- **Recording data**

The first stage of any investigation is to clearly identify the problems to be solved or questions to be asked, and to have a sound idea of how the **data** collected will be analysed. This allows the experiment or survey to be set up to satisfy the aims of the study and thus to ensure that the subsequent conclusions are based on evidence and not supposition. When beginning a piece of research, it is important to approach it systematically. To get the most out of any research, make sure you identify the practicalities involved, and work within the available resources of time, space and equipment, as well as acknowledging any safety and legal issues (see Box 1.1).

Box 1.1 *Safety and legal issues*

Before embarking on any type of research, be it field or laboratory based, there are certain safety, legal and practical aspects which should be considered. In England, Scotland and Wales, under the Health and Safety at Work Act (1974), it is important to take reasonable care of the health and safety of all concerned. You should produce a risk assessment for the research, incorporating all risks involved in the experiments or surveys you intend to perform, including any regulations concerning substances with which you will come into contact, such as the Control of Substances Hazardous to Health Regulations 1988 (COSHH). Any local or institutional health and safety regulations must also be adhered to. Always operate within the law, obtaining permission to work and take samples in your study area. See Nichols (1990) and Pigott and Watts (1996) for further information.

The flowchart in Figure 1.1 identifies the stages involved in the successful design and implementation of research. Projects start with a question, and the simpler this question the better. Even the most apparently simple of questions opens up a realm of

possibilities for research design. This is best illustrated by an example. If we were interested in whether there was any difference in pollutant levels in produce from allotments at different distances from major roads, our first task would be to narrow down the aims. Since there are several possible pollutants and perhaps a number of different crops, we could, for example, first narrow the investigation to look at lead accumulation in cabbage plants. Thus, a working title might be 'Lead accumulation in cabbages grown in allotments at different distances from major roads'. The next stage is to refine the question further into a testable **hypothesis**; this is where the statistical process starts. The question 'Is there a difference between lead accumulated in cabbages from allotments near to major roads and those far from roads?' is a testable hypothesis, and can be broadly defined as a difference question.

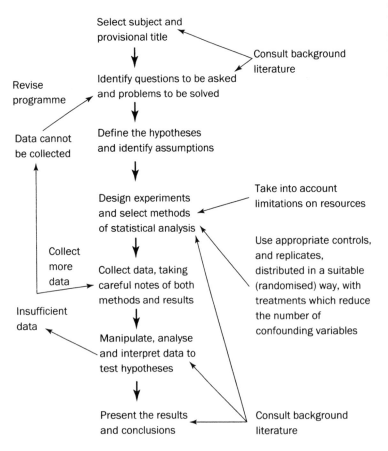

Figure 1.1 *Sequence of procedures in the planning and implementation of research*

Another testable hypothesis that could have developed from the aim of the study takes the form of a relationship question. By taking cabbages at several distances away from a road, we could test whether the lead content of cabbages increases or decreases with distance from the road. The general form of the hypothesis (in this example, difference or relationship), and the exact sampling method determine the family of **statistical tests** we later use.

To answer either of these questions, we need to know two pieces of information about any given cabbage: its lead content and its position relative to a road (using the categories of either near or far for the difference question, or the actual distance in metres of the allotments from the roads for the relationship question). These pieces of information are known as **variables**.

Variables

When we ask a difference question (i.e. is there a difference between lead accumulated in cabbages from allotments near major roads and those far from roads?), for each individual cabbage we would record two variables: distance from the road (either near or far) and the lead content. The first of these variables (distance) is set by the investigator in the design of the survey because allotments are chosen at certain positions relative to the road (near and far) and is therefore known as a **fixed variable**. The second variable (lead content) is free to vary, and is known as a **measured variable**. Variables may also be classified according to whether they are dependent or independent. The lead content is termed the **dependent variable**, since it may depend on the value of the fixed variable (i.e. lead content may depend on whether allotments are near or far from roads). The distance from the road is then known as the **independent variable** (because the distance an allotment is from a road cannot depend on the lead content of the cabbages growing there). We also need to be aware of any **confounding variables** (i.e. those independent variables that we are not directly interested in but that vary in a similar way to the ones we are measuring). For example, traffic levels may also affect lead content. Proximity to the road may be more or less important than level of traffic, but, without taking traffic into account, we may not be able to separate the effects.

Populations, samples and individuals

To test the hypothesis that there is a difference between cabbages near and far from a road, we could not simply take a lead measurement from a single cabbage from an allotment near to a road and another measurement from a cabbage in an allotment further away. This is because differences between individual plants, for example in terms of their biology or exact position in the allotment, will cause variation in the way in which they accumulate lead. This is a key statistical point: the characteristics we are trying to measure are inherently variable. In statistical jargon, all cabbages near to the road are individuals in a statistical **population** (the population in this case is defined as cabbages in allotments near to major roads). The individual members of the population vary in their values of the characteristic measured (in this case lead content). The exact level of variation of this characteristic usually remains unknown to the investigator, since it would be impractical, uneconomic and/or impossible to measure all individuals in a population. The way we get around this problem is to take a **sample** (a group of representative individuals from the population) from which to estimate the extent of variation in the characteristic.

So far, individuals and populations have been easy to visualise; in this example, each cabbage is an individual from a population of all cabbages that could be investigated. Similarly, in a survey of levels of radon gas in homes, each house would be a statistical individual (from a population consisting of all houses in the sampling area). However, visualising the statistical individual and/or population is not always so simple. In many experiments or surveys we take counts within a certain area, volume or time period. If we examined soil moisture content by taking soil cores, the statistical individual is the core (within which we could measure the moisture content), and the statistical population comprises all possible cores that could have been taken in the sampling area.

Observations and manipulations

There is a further distinction between project designs which is worth introducing at this point. The results of the difference test on the lead content of cabbages near to and far from a major road would tell us whether there was a difference in lead content. However, it would not definitively tell us whether any difference was due to the distance from the road or whether it was because of some (unmeasured) variable which varies systematically with the position of the allotments in relation to the road. Although in this case it would seem likely that proximity to the road would be a major causative factor (particularly if we had taken the potentially confounding variable of traffic level into account), there are other situations where cause and effect can be wrongly assumed. A major problem with **observational surveys** such as these is not being able to take into account all potentially important variables. The only way to truly demonstrate cause and effect is to set up a controlled **manipulative experiment**. In a manipulative experiment the investigator controls most of the variables (these are therefore the fixed variables), and only the variable(s) of interest are allowed to change (measured variables). The different values of the fixed variables are often called **treatments**. This terminology originates from agricultural statistics, the science for which many of the statistical techniques we use today were developed. In an agricultural context, crops could be treated with a chemical to examine the effect on crop growth (the measured variable), and the various doses of the chemical would be the statistical treatments (the fixed variable).

To find if a treatment has an effect on the measured variable, the same measurements are also made in the absence of any treatment. This is known as a **control**. Controls are set up under the same conditions as the treatments so that they differ only in the absence of treatment. For example, to find out whether adding a fertiliser to a derelict site influences plant colonisation, we would compare areas treated with fertiliser to a control where no fertiliser was applied. However, since fertiliser is applied in an aqueous solution, it is possible that the water in the solution also affects plant growth. Therefore, to find the effect of fertiliser alone, the correct procedure is to add an equivalent amount of water to the control areas.

Individuals are randomly assigned to treatments so that there is no difference between individuals at the start of the experiment. For example, if we were interested in which of two species of tree grew best on contaminated land, then we could set up a series of experimental plots comprising the materials which are found on such sites, and which were as similar as possible to each other. Half of the plots could be planted with one species of tree and the remainder with the other, with individual trees being randomly allocated to plots, so that there would be no systematic difference in the environment.

The disadvantage with manipulative experiments in environmental research is that they typically take place in experimental fields or laboratories where the conditions are so tightly controlled that the applicability of the results to the real world can be hard to assess. In contrast, many environmental investigations are observational, for example, surveys of the levels of pollutants in water, soil or air. These surveys cannot normally demonstrate cause and effect, although they can be strongly suggestive and useful in providing a source of hypotheses which may be tested using manipulative experiments. However, in many cases the ideal manipulative experiment to confirm such cause and effects is not feasible. A classic example is the causal relationship between smoking and lung cancer in humans. The proper experiment would be to assign people at birth to each of two treatments (smoking and non-smoking), forcing the smokers group to smoke and not allowing the non-smokers access to cigarettes, while keeping all other factors the same. Since we clearly could not carry out such an experiment, it is impossible to prove that smoking causes lung cancer in humans. However, a combination of manipulative experiments on other animals, and observational surveys that take into account a great many variables and examine large samples of people who do and do not smoke, have produced a large amount of evidence which supports the hypothesis that smoking can lead to lung cancer.

Generally in environmental research, manipulative experiments are most useful if kept simple, since the more fixed variables and the more treatments within each fixed variable that are incorporated, the more difficult the implementation and the more complex subsequent analysis and interpretation. In systems which are relatively easy to manipulate (e.g. in agricultural or psychological research), designing experiments with a large number of different treatments for several fixed variables can lead to sophisticated (and economic) research (i.e. enabling several questions to be answered in a single experiment). However, these experimental designs also require sophisticated analysis techniques. In general, whatever type of experiment or survey, the most elegant research providing the most meaningful results tends to be the result of simple, well-thought-out designs.

Sampling

As we have seen, when we measure some aspect of a population, it is no good simply taking a single reading because it may be unrepresentative of the population as a whole. Therefore, in any sort of experiment or survey we need a method for selecting items for study. The way in which we do so is crucial, because only individuals drawn randomly from a population can be considered to represent the population. If there is any bias in the selection of the items to measure, then our sample will not be representative. In our cabbage example, without a strategy planned prior to obtaining the cabbages, it is easy to imagine a researcher subconsciously selecting the plumpest, healthiest-looking individuals. Samples must be taken using criteria set up in advance of the survey and should cover all the variation present. Sometimes it is difficult to obtain true random samples. For example, it is often hard to obtain an unbiased sample of animal populations, because if you have to trap the animal, the individuals caught are actually from a population of animals that are more susceptible to being trapped. Similarly, it is very difficult to obtain a truly random sample of people (it is one thing to randomly select a name; it is another to track down and persuade the individual to take part in the study). These sampling problems may be unavoidable. The key point to make here is that you must describe exactly how you obtain samples when you write up a report, in order that others can interpret your results accordingly.

If we were to sample random cabbages in an allotment, we could visualise the allotment as a grid (see Figure 1.2). Anywhere on this grid has a coordinate (map reference). Three ways to sample the allotment are presented in Figure 1.2. In Figure 1.2a, cabbages are taken at evenly spaced intervals. This **systematic** approach will give a random sample of cabbages only if cabbages are randomly situated in the field in the first place. It is possible, however, that the distribution of cabbages varies systematically over the allotment, for example cabbages every metre apart may be smaller than those in between. While this seems unlikely in this example, we cannot rule it out and therefore cannot know whether the sample of cabbages is truly random. An alternative method of sampling (Figure 1.2b) is to take random coordinates within the grid. The coordinates could be generated from random numbers obtained from a pocket calculator or from statistics tables (e.g. Neave, 1995). The disadvantage of this truly **random sampling** method is that sometimes, purely by chance, a particular area of the grid may be underrepresented. For example, there are not many points in the bottom right-hand corner of Figure 1.2b. If this section in the allotment was different in any way (e.g. if it had a tendency to be wetter), then the sample would again not be representative of the population. For spatial sampling, the best strategy is often a combination of the two: the **stratified random** method (Figure 1.2c), where within each square of the grid an individual is taken from a random coordinate. Similar strategies apply to sampling in other

(a) Systematic   (b) Random   (c) Stratified random

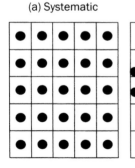
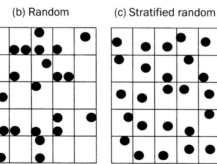

Figure 1.2 Sampling strategies

situations: for example, if taking measurements of pollution from cars, the vehicles may be selected by using every tenth car (systematic) or using random numbers to select which cars to sample (random).

Block designs

A stratified random approach (Figure 1.2c) is particularly useful when the sampling area is not uniform (e.g. a meadow with different conditions at each end). When the statistical population contains subgroups (e.g. an area of study with patches of wetter ground, areas differing in soil type, or different types of people using a recycling facility on different days), stratified random sampling can be based around these subgroups. The subgroups are often known as **blocks**. Blocks are fixed variables which we may not be directly interested in, but which we can take into consideration in the design and analysis of an experiment or survey. Usually, each block would have the same number of individuals sampled from within it. A block can be:

- **Spatial**. For example, several sections of land within which each of a series of treatments takes place (in agricultural terms, this is what blocks were originally designed for), or several different cities where we are interested in differences between the atmospheric pollution associated with the centres and suburbs of each city.
- **Temporal**. For example, periods of time within which people are surveyed at random (morning, afternoon and evening), or, as in our cabbage example, the order in which measurements of lead content are taken from different categories of the fixed variable (near and far from a road). If all the measurements on cabbages from the allotments near to roads were performed first, followed by all the ones far from roads, a difference could emerge that was simply due to bias. For example, the researcher, after an enthusiastic start, could become bored with the repetitive laboratory procedure, making the afternoon measurements less reliable. Alternatively, the cabbages could deteriorate over time, giving the later ones lower readings.

The block design in manipulative experiments

In manipulative experiments, specific treatments should be separated in such a way that adjacent treatments are not always the same; using block designs provides a method of doing this. For example, in an experiment to determine the impact of various pesticides on the beneficial insect community, pesticide A should not always be used next to pesticide B (see Figure 1.3a–b). One way of ensuring that pairs of treatments are not always together is to use a **randomised block** design where each treatment is allocated at random to a position on the row (Figure 1.3c). Alternatively, a random **Latin square** could be used, where each treatment occurs once in each row

| (a) Systematic design | | | | | | (b) Systematic offset design | | | | | | (c) Randomised block design | | | | | | (d) Random Latin square | | | | |

(a) Systematic design

A	A	A	A	A
B	B	B	B	B
C	C	C	C	C
D	D	D	D	D
E	E	E	E	E

(b) Systematic offset design

A	B	C	D	E
D	E	A	B	C
B	C	D	E	A
E	A	B	C	D
C	D	E	A	B

(c) Randomised block design

B	C	E	D	A
D	E	B	C	A
B	E	A	C	D
C	B	E	D	A
C	D	E	A	B

(d) Random Latin square

E	A	D	B	C
D	B	E	C	A
A	C	B	D	E
C	D	A	E	B
B	E	C	A	D

Each row has a different treatment, while each column contains one of each treatment

Each row and column contains one of each treatment

Each row contains one of each treatment in a random order

Each row and column contains one of each treatment in a random order

Figure 1.3 *Possible layouts of experimental blocks*

and once in each column of the square, so that the sequence of treatments by row or column is random (Figure 1.3d). As the dimensions of the square increase, the number of possible ways of arranging treatments also increases. If we use one of the random designs (Figures 1.3c–d), we can be reasonably confident that some environmental variable, either foreseen or unforseen, will not bias the results. If we want to actually quantify the effect of the block, it can be incorporated into the analysis; these more advanced statistical techniques are mentioned in Chapter 7. When laying out plots like these, it is worth considering the effects of edges. Environmental conditions (e.g. microclimate) may be very different at the edge compared to the centre of a series of experimental plots. One way of reducing this edge effect is to incorporate a buffer zone surrounding the plots into the design, where treatments are still applied, but measurements are not taken.

These randomised block designs are commonly used in field experiments; however, they are also applicable to the way in which many other experiments are designed. For example, the squares in Figure 1.3 could represent different shelves or positions of experimental trays in a culture room or greenhouse, which could be subject to different conditions in the same way that different parts of a field may experience different conditions. The squares could also represent different times of day (with each row being a single day), the blocks being the order in which individuals from different treatments (A, B, C, D and E) are measured.

Sampling people

The methods for obtaining a representative group of people follow the same general sampling rules. The exact sampling method used is influenced by the target audience.

For example, if we wanted to obtain information regarding the energy efficiency practices of local industries, then individuals (i.e. each industry or factory) are relatively rare and a postal or telephone questionnaire to a representative from each of them may be appropriate. Eliciting the responses of local residents to changes in public transport facilities will probably require a postal or doorstep interview, where local addresses could be sampled randomly. In contrast, where users of a facility such as a recycling site are to be surveyed, face-to-face interviews on-site will probably be best.

First consider whether a random, systematic or stratified approach will be used. If all potential respondents can be surveyed, then sampling the total population is the most reliable method (this is a form of systematic sampling). For large populations, a random sampling approach (e.g. selecting addresses by using random numbers against lists in directories) or a systematic approach (e.g. taking every fifth or tenth address) will help protect against biased selection. It may be that roughly equal numbers of subgroups are required (males and females or young and old respondents). In this case, although selection can still be random, surveying could stop once a particular quota of a subgroup has been reached (**quota sampling**). Another method which is sometimes used is that of **accidental sampling**, for example where the first 100 people to be intercepted are approached.

Cluster sampling can be useful when the population to be sampled occurs in groups. For example, if you wished to examine the number of British children who are members of an environmental organisation, the ideal (but impractical) way forward would be to randomly select a sample from all children in Britain. To make this more manageable, you could use the fact that children tend to be clustered in schools and randomly select a sample of British schools to work on and then subsample the children within each school.

The technique of **snowballing** relies on each selected item generating several more (e.g. in questionnaire surveys, respondents could be asked to forward a copy of the questions to two or three other people). Since the individuals generated by this method are not random, it is not an ideal way to sample the population (although sometimes it may be the only practical option).

Once people have been approached to take part in a study, another issue is how many respond or agree to take part. It is ideal if 80% or more of the individuals approached agree to take part in the survey. The problem arises if there are too many refusals, because the sample is then biased towards people who, for whatever reason, have a particular interest in taking part. Postal surveys are particularly prone to low response rates; it is not uncommon to find that as few as 10% return the questionnaires. Targeting a named person or a position in an organisation (e.g. the planning officer), pre-warning of the arrival of the questionnaire and following it up by letter or telephone are ways in which response rates may be improved. Explanatory information regarding the reasons for the survey may also help. The response rate (i.e. the percentage of people agreeing to take part) should be stated in the methods section of a report, so that others can critically evaluate the study.

Questionnaire design

When eliciting information about human activities, or the opinions and attitudes of a section of the population to a particular issue, then some kind of interview technique is appropriate. Often this comprises a series of questions which are either filled in by the interviewee (self-administered) or recorded in a structured way by an interviewer. First decide on the method of presenting the questions. Using self-administered questionnaires, the respondent has the opportunity to read the entire questionnaire and to take time to consider his or her answers. However, this means that, compared to a structural interview, there is less spontaneity and a greater chance of a reply which is either tailored to a particular image (however subconsciously) or which may be assisted by others. There may also be a low return rate from self-administered questionnaires (this poses problems as to how much the conclusions are relevant to the general population – see the earlier discussion on sampling people). If the interviewer presents the questions, not only are answers spontaneous (thus hopefully representing the individuals' true feelings, interests or opinions), but there is an opportunity to clear up ambiguities and clarify the respondents' answers. However, this is much more time-consuming than using self-administered questionnaires. Also, where questions are very personal or touch on controversial subjects, face-to-face surveys may intimidate some respondents.

How the questions are asked varies from topic to topic, both between and within questionnaires. Questions may be open or closed: **open questions** ask for the respondents' comments without restricting the possible answers; **closed questions** limit the respondent to one of a number of responses listed as possibilities, although sometimes an additional open response may be useful. Examples of open and closed questions are given in Table 1.1. Open questions are more difficult to analyse and often are simply examined in terms of the frequency of responses of a similar nature. However, they can be useful in finding out what truly interests the respondents. It is often useful to use an open question at the end of a questionnaire to pick up items you did not think to include. The amount of information gained from questions such as those illustrated in Table 1.1 can be increased if respondents are asked to list the items in order of preference or importance. This enables each item to be given a score (e.g. with the item of lowest importance being given a score of 1, that with the next lowest importance a score of 2, and so on).

It is essential that questions are not ambiguous or biased. Take, for example, the question: 'do you agree that the irresponsible behaviour of the company concerned has severely damaged the local environment?' This is a leading question, expecting the answer 'yes'. Whatever your opinions, you should ensure that the questions are presented in an unbiased way. This is also important in face-to-face interviews where the form of words and emphasis placed on them should not vary between interviews. Pilot studies, where the interview or questionnaire is tried on a few people first, should be carried out to identify confusion or ambiguity in the questions.

Table 1.1 *Comparison of open and closed questions to examine user attitudes to the facilities provided in a country park*

Open question	Please list below the facilities that you feel should be provided at this park:

Closed question	Please tick below which of the following facilities you feel should be provided at this park:
	• visitor centre
	• hides
	• toilets
	• car parking
	• gravel paths
	• signposting
	• information boards

Closed question with an open response	Please tick below which of the following facilities you feel should be provided at this park:
	• visitor centre
	• hides
	• toilets
	• car parking
	• gravel paths
	• signposting
	• information boards
	• other (please specify) . . . .

The layout and length of the questionnaire will also depend upon the subject of the survey. Whilst it is important to cover the main points, try not to ask too many questions and keep them relevant to the topic of the research. Large unwieldy questionnaires not only put respondents off, but also require lengthy data entry and analysis. Make good use of the space available to allow the respondent to feel at ease with the format, and avoid a cramped sheet where questions may be overlooked. It is often useful to ask for factual information about the respondent first and then move into the major aspects of the survey.

The example questionnaire in Figure 1.4a is ambiguous and provides little guidance to the respondent. Some users may not fill in a questionnaire which does not have an obvious aim. Others may be put off by the relatively poor layout and the lack of direction. For a survey of this nature it is worth considering whether you need the respondent's name and address. This is especially important if a sensitive issue is being raised, where it is also important to stress confidentiality and/or anonymity.

(a) Unclear, leading and ambiguous example

(b) Clearer and more helpful example

```
Questionnaire for facility users

1.   What is your name?.....................

2.   What is your address?..................
     ........................................
     ........................................

3.   What is your age?......................

4.   How frequently do you use this facility?
     ........................................
     ........................................

5.a  Do you agree that the opening hours for
     this facility are inadequate?
     ........................................
     ........................................

5.b  If you find the opening hours for this
     facility inadequate, why is this?
     ........................................
     ........................................

5.c  If you find the opening hours for this
     facility inadequate, would you use the
     facility more frequently if the
     opening hours were altered?
     ........................................
     ........................................

etc.
```

```
We are surveying users of this facility to
ensure it runs as efficiently as possible.
Please help by filling in this questionnaire,
ticking the boxes as appropriate; your
answers will help us improve the facility.
Thank you for your help.

1.   Age: under 18☐  19-30☐  31-50☐  over 50☐

2.   How frequently do you use this facility?
     At least weekly                       ☐
     At least monthly                      ☐
     Fewer than 12 times per year ☐
     Never before                          ☐

3.a  Do you find the opening hours
     Adequate    ☐ If ticked go to Q.4
     Inadequate ☐ If ticked go to Q.3.b

3.b  Are the opening hours:
     Too short                             ☐
     Not early enough                      ☐
     Not late enough                       ☐
     Other (please specify).............

3.c  Would you use the facility more
     frequently if the opening hours were
     altered to accommodate the problems you
     identified in Q.3.b?
     Yes                                   ☐
     No                                    ☐
     Don't know                            ☐

etc.
```

Figure 1.4 *Example layouts for questionnaires*

Similarly, providing a range of options for age may encourage responses from those who dislike giving their actual age. It would also be difficult to analyse the data obtained from the questions in Figure 1.4a because the reader is not guided as to the form of answer required (e.g. for question 4 some may answer 'occasionally'). In contrast, Figure 1.4b explains the aims of the survey, and the questions are relevant and structured to avoid ambiguity. A number of texts (especially in the social sciences) describe in more detail the principles of questionnaire design (e.g. de Vaus, 1996).

Semi-structured and unstructured interviews

Less structured interviews are usually employed to gain a more in-depth perspective than that obtained from questionnaires. For example, if we were examining the role of a local authority in the planning process, it might be useful to interview a variety of participants in a recent planning application to identify how successful they felt the process had been. Here, we could take a broad approach where people were individually asked about their role and their various responses used to stimulate further questions (**unstructured**). Alternatively, a series of previously identified questions could be asked, with supplementary questions being stimulated by the responses (**semi-structured**). In either case quantitative analysis is difficult. Frequently such investigations are written up as case studies rather than using data analysis – see texts such as Robinson (1998) and Lindsay (1997) for further details.

If such interviews are recorded or written up in full, then aspects of the responses could be quantified using a technique such as **content analysis**. Here the number of positive and negative statements made can be tallied, or the depth of feeling on a subject evaluated by the emphasis placed on it by the respondent. Content analysis can also be used to examine the way in which written articles (from newspapers, journals, company documents, etc.) report particular facts: whether the words used are positive or negative, whether the items are reported in a prominent position or hidden, how many words are used to describe the item, etc. – see texts such as Weber (1990) for further details. Once data have been assembled in numerical terms (length of article, ratio of positive statements to negative ones, etc.) they may be analysed in the same way as other numerical data.

Group interviews (also called focus groups) often produce a wider-ranging discussion than would have resulted from individual interviews. They also enable several people to be interviewed in a relatively short period of time. Care needs to be taken in the selection of interviewees: using a mixture of ages, ethnic backgrounds, sex, etc. may help stimulate debate, although a group that is too diverse may form subgroups. Large groups (over 12) may suffer from fragmentation into subgroups and/or the lack of participation by quieter members. Small groups (under 6) may require more input from the researcher to maintain the discussion. The researcher may wish to direct the discussion (either formally or informally) or simply be present as an observer. Group interviews may also be a good method of focusing on potential questions to be used in a larger interview or questionnaire study.

Sample size

The number of individuals in a sample is called the sample size. The sample size has to be sufficiently large to cover the inherent variability in the characteristic measured. The exact size of the sample required depends on knowledge of this variability (which is usually difficult at the beginning of the experiment or survey), or may depend on the resources available. Although it is usually advantageous to sample as large a proportion of the population as possible, this will be limited by economic and environmental considerations. Running a pilot study will give an indication of the magnitude of the inherent variation, and hence the likely sampling intensity needed, and may also allow the final experiment or survey to be modified to avoid unforseen problems. To make best use of resources, and to have the best chance of covering inherent variability, it is usually preferable to have sample sizes as close to equal as possible. Some statistical techniques actually require equal sample sizes: one more reason why you need to know how to analyse the data before collection.

With questionnaires, the number of respondents surveyed will depend on the resources available and the size of the population. As a rule of thumb, the minimum number which can be analysed effectively is about 30 for each subgroup surveyed. Thus, if the final analysis requires the data to be broken down into four age classes examining male and female respondents separately, then the total number of questionnaires should be at least 240 (30 individuals in each subgroup × 4 age groups × 2 sexes). Of course, this does not take account of those questionnaires which are not returned or those only partially completed. A pilot study will allow you to assess what the response is likely to be, and adjust the number of questionnaires sent out.

Independence of data

It is important to obtain unbiased samples. That is, the selection of one individual should not influence the likelihood of including any other individual, and each individual should be independent of each of the others. Imagine an experiment where plants were grown on contaminated and on clean soil in order to compare the condition of their leaves. Although the aim was to inspect a standard leaf (e.g. the eldest) from each of 50 plants from each soil type (100 plants in total), the plants on the contaminated soil grew so poorly that there were only 30 plants. It might be tempting under these circumstances to obtain two leaves from some of the plants on the contaminated soil to make up the numbers. However, it would be wrong to do this, because leaves sampled from the same plant do not represent independent data points. The two leaves would share more characteristics with each other than with the other available plants (e.g. they would be more likely to be in a similar condition to each other). In this case, to achieve **independence**, it would have been better to use two samples of unequal size. A few tests require equally sized samples; in this case it would be best to reduce the number samples taken from the plants on uncontaminated soil.

There are occasions where non-independent data are deliberately used as part of the design of an experiment. For example, if a measurement was made on the same individual before and after treatment, then each individual would have two **matched** data points. Similarly, if we had, say, ten cities for an investigation into urban atmospheric pollution, then by recording the pollutant levels at the centres and edges of each city, we would have two matched data points for each city. This is a paired or matched group design, and would be analysed with a particular type of statistical test (paired tests are explained in Chapter 4). However, it would still be important for individuals (i.e. cities) to be independent of each other.

Sometimes in environmental experiments it is hard to avoid non-independence of data. For example, if we wanted to compare the number of beneficial insects inhabiting fields sprayed with pesticides with those which are not, it would be ideal to use several fields (say 20) in order to have replicate plots (i.e. 10) of each

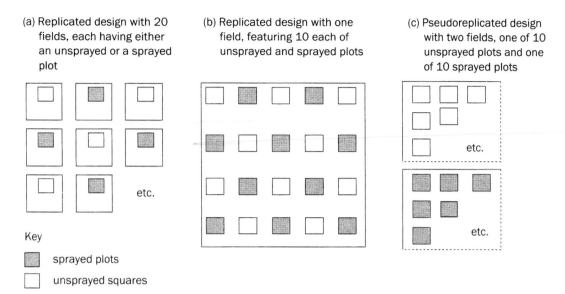

(a) Replicated design with 20 fields, each having either an unsprayed or a sprayed plot

(b) Replicated design with one field, featuring 10 each of unsprayed and sprayed plots

(c) Pseudoreplicated design with two fields, one of 10 unsprayed plots and one of 10 sprayed plots

Key

▨ sprayed plots

☐ unsprayed squares

Figure 1.5 *Comparison of replicated and pseudoreplicated experimental designs*

type (Figure 1.5a). If study sites are limited, an alternative would be to use several different blocks of both the treatment and control in one field (Figure 1.5b). However, a common mistake is to rely on one example of each type (e.g. one field which has been sprayed and one which has not), with several plots (say ten) being surveyed within each (Figure 1.5c). The outcome in terms of the number of data points would be the same (i.e. ten measurements from sprayed samples and ten from unsprayed). However, the latter method is false replication (**pseudoreplication**), because it does not take into account differences between the fields: this sort of design should be avoided where possible. Unfortunately, in practical terms this is sometimes the only realistic way forward, and many studies have used similar designs. If a study has used a pseudoreplicated design, we need to interpret the data accordingly; the results compare these two fields only. Any differences found are not necessarily the result of application of pesticide, as there could be other differences between the two fields which, because we have not truly replicated the work, we cannot measure. Further experiments will be needed to test the general implications of the work.

Sources of error

Any sample is inherently variable. Ideally, this variability should reflect the true variability of the population. Unfortunately there are a number of sources of error which may add to, or mask the nature of, this variability. Many of these errors can be avoided by some care and attention to detail, as explained in Table 1.2.

Table 1.2 *Sources of experimental error*

Type	Source	Method of reducing the error
Human errors	Carelessness (e.g. reading the scale incorrectly on an instrument, or incorrectly recording data items).	Take care and pay attention to detail, check data for obvious errors in scale, position of decimal point, etc.
	Lack of practice of the technique (e.g. not recognising the difference between two species).	Make sure that all researchers on the project are properly trained before data collection commences.
	Increasing fatigue throughout a day of data collection, causing increased errors in later measurements.	Randomise the order in which sites are visited, or measurements taken.
Instrumentation errors	Limitations of equipment (e.g. trying to measure at levels of precision beyond the machine's capabilities).	Know the limitations of any equipment and/or technique being employed and keep within its capabilities; do not attempt to estimate values which fall between marks on analogue scales.
	Presentation of data at a greater degree of precision than was measured (e.g. recording more decimal places than is justified, especially when data have been transformed from one unit to another).	Use the appropriate level of precision for the measurement being taken.
Systematic errors	Bias in measurement (e.g. due to faulty or badly calibrated equipment – pH meters calibrated to pH 4 may not be accurate at high alkalinity).	Check equipment and calibrate properly for the range of values you expect to take; ideally check values using more than one technique.
	Bias of researcher (e.g. when samples are taken because of some conscious or unconscious preference).	Use random sampling techniques where possible.
	Decay of specimens during storage prior to data collection (e.g. water, soil or living specimens can change in physical, chemical and biological properties).	Randomise the order in which specimens from different categories or treatments are measured.
Unrepresentative samples	Not all of variation covered (e.g. if different soil types are present when monitoring vegetation cover).	Use stratified random sampling where appropriate.
Uncontrolled factors	Changes which cannot be anticipated or managed (e.g. climatic differences between sampling days).	Monitor and build into data analysis and interpretation where possible.

Confounding variables (those which alter at the same time as the measured independent variables) can be assessed and minimised. For example, where different researchers gather data on separate sites, if a difference is found between sites, it may be a result of the level of researcher effort or skill. This problem would be avoided by ensuring that where possible all researchers survey each site for the same amount of time (where necessary, the recorder can be built into the analysis as an extra variable).

Counts and records of objectively based values are accurate when made carefully. Measurements on infinite scales such as length, on the other hand, are reliant on the accuracy of the recorder and the measuring device. Where a rule is used which has divisions in 1 mm units, then each measurement is accurate to 0.5 mm. That is, a length of 4 mm is in reality between 3.5 and 4.5 mm. If length is measured using a calliper with divisions at 0.1 mm then the readings are accurate to 0.05 mm. The degree of precision used to measure variables on infinite scales is usually selected so that there are between 30 and 300 units between the smallest and largest values measured. Both accuracy and precision need to be maximised: you can be accurate in measuring the number of plants found in a 0.5 m² quadrat, but not precise if an average of several attempts at the same quadrat arrives at the correct figure, while having a large variation in your counts. Conversely, if you arrive at the same number of plants on each of your attempts, but in reality your results are always an underestimate, perhaps because you invariably miss several small plants, you are being precise, but inaccurate (see Box 1.2).

Box 1.2 *Comparison of precision and accuracy*

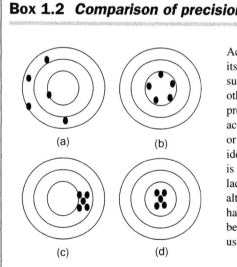

(a)   (b)

(c)   (d)

Accuracy is the closeness of a measurement to its true value, while precision is the closeness of successive measurements of an item to each other. Measurements may be neither accurate nor precise (a), may lack precision but be relatively accurate (b), may be precise but inaccurate (c), or be both accurate and precise (d). Although the ideal situation is for both to apply (d), accuracy is usually more important than precision, since a lack of accuracy may be due to bias, whereas although lack of precision (b) may make it harder to identify differences or relationships between samples, it may be compensated for by using a large sample size.

Data types

Within a sample, the characters to be measured are termed variables (e.g. length, colour, number of items, time, mass). The variables recorded may be:

- categories (e.g. sex, type of environment);
- ranks (e.g. the strength of opinion regarding a particular topic, complexity of environment);
- counts (e.g. number of species, numbers of people, numbers of plants germinating);
- physical measurements (e.g. pH, water content, length, width).

We need to understand the differences between the data types because these affect the statistical tests required. It is sometimes possible to record the same variables in a number of ways. For instance, assessment of plant colour could be made by: measuring the wavelengths of transmitted light from the surface under standardised conditions; comparing the colour with a series of standard charts and classifying each according to the nearest match; ranking plants from darkest to lightest; or categorising each plant according to specified criteria (e.g. yellow, green, brown, orange). In each case the type of data obtained is different and must be handled in a specific way. In most cases, the more numerical the data, the more rigorous the analysis that can be used, with physical measurements being the most numerical and categories the least. There are three major scales of data (see Box 1.3) which basically equate to category data (**nominal** data which identify broad characteristics about the items such as colour or sex), ranked data (**ordinal** data which label the position relative to other values such as large, medium or small), and measurement data (**interval** and **ratio** data which give precise information about the magnitude of particular characteristics such as numbers of items, length, mass and area).

Variables measured on interval or ratio scales may be **discrete** (discontinuous) or **continuous**. Discrete variables are measured in units with distinct gaps between adjacent units (e.g. are assigned values in whole numbers), whereas continuous variables are placed on a scale with an infinite range of points. The number of landfill sites in a region, the number of species of small mammals in a woodland, and the number of people using a footpath are all discrete measurements (you cannot say there are 2.5 landfill sites in a region). Note that although discrete variables are always measured in whole numbers, subsequent derivations (e.g. means) do not have to be in integers (e.g. there could be a mean of 2.5 landfill sites per region). The concentration of a pollutant, the wind speed, and the area of a nature reserve are all continuous measurements.

Another distinction between data types, which is important for some statistical methods, is whether variables are **derived** or measured directly. An example of a derived variable is a proportion or rate calculated from other measurements (e.g. the number of individuals belonging to one plant species may be converted to a percentage by dividing it by the total number of plants of all species found and

Box 1.3 *Data scales*

Nominal (or categorical) data are placed in pigeon-holes (e.g. type of motor vehicle – car, minibus, coach; or sex – male, female). Such variables are often used to separate samples for analysis (e.g. mean pollutant levels from different types of vehicle can be compared to see whether they differ in the amount of lead they emit). They may also be used for frequency analysis (e.g. examining the number of males or females who are members of a particular environmental organisation). The categories are mutually exclusive (i.e. a measurement must not be able to fit into more than one group) and it is not possible to identify one item as being larger than another. Although numbers can be used to code each category (e.g. male = 1, female = 2), these are purely labels and have no value. It is better to use names (e.g. male, female) or non-numerical labels (e.g. M, F) to avoid confusion (although some statistics programs do not accommodate this).

Ordinal (or ranked) data are on a scale with a defined direction (i.e. one point can be described as larger than another), but with distances between adjacent points not necessarily equal. For example, when plant abundance is recorded on the DAFOR scale (Dominant, Abundant, Frequent, Occasional, Rare), although abundant is larger than occasional, we cannot say that it is, say, twice as large. For analysis, ordinal data are given ranked values (e.g. rare = 1 and dominant = 5). Despite the fact that those ranked 1 are lower than those ranked 5, they are not 5 times lower. Data allocated to size classes are also ordinal (e.g. 0–9 mm, 10–19 mm, 20–29 mm, etc.). Ordinal data can be used in many types of analysis, but care needs to be taken to use appropriate tests.

Interval and ratio data are on measurement scales with a defined direction (i.e. it is possible to state that one point is greater than another) and with measurable intervals between values. Interval data have no absolute zero value, so we cannot state that an item with a value of 20 is twice one with a value of 10. For example, measurements of vehicle exhaust emissions collected on the 20th of October were not taken twice as recently as those monitored on the 10th. Temperature on the Celsius scale is also measured on an interval scale: 20°C is not twice as hot as 10°C since the zero point is arbitrary and not absolute (i.e. there are possible negative values). Ratio data have absolute zero values, so a value of 20 is twice as big as a value of 10, and negative values are impossible (e.g. –3 plants and –6 km are nonsensical). Length, width, mass, and temperature on the kelvin scale all lie on ratio scales. Both interval and ratio data are analysed in the same way, using the most powerful statistical tests.

---

multiplying by 100). These data would still be derived even if they were measured directly (e.g. estimates may be made of the percentage cover of a species within a quadrat). Other examples of derived variables include the ratio of site length to breadth as a measure of shape, the proportion of a country's energy which is generated using wind power, and the rate of flow of a stream. Derived measurements may need special manipulations (transformations) before analysis (see Chapter 3).

Measurement data give us the most information about a system, whilst nominal data give us the least. It is possible to downgrade data if required. So, measurements of length could be converted to an ordinal scale by allocating them to fixed ranges such as 0–4.9 mm, 5–9.9 mm, 10–14.9 mm, 15–19.9 mm, 20–24.9 mm (note that these do not overlap, allowing any measurement within the range used to be placed in one category alone). Such conversions reduce the detail in the data and should only be used when necessary, for example to combine precise measurements on a ratio scale into broader classes in order to display frequencies of occurrence (see Chapter 2). In general, it is a good idea to collect data as precisely as possible since more powerful tests tend to use more detailed scales of measurement. You can always downgrade data later, but you cannot reclaim lost detail if you record initially at too coarse a scale.

In questionnaire design there may be sound reasons for not collecting the most detailed data. There are some (often more personal) questions where a response is more likely if the respondent can select from a band of responses rather than enter the actual figure (e.g. salary and age). There are other questions, although not personal, for which you cannot expect accurate answers, simply because it is too difficult for the respondent to remember in detail. The number of options used for rankable questions should be reasonable for the topic. So, asking how frequently someone visits a local park could have ranked responses of: at least once per day; at least once per week; at least once per month; at least once per year; less than once a year; and never. Whereas asking about frequency of visits to a recycling depot would probably have a different time frame such as: at least once per week; at least once per month; less than once a month; and never.

Attitudes and opinions are often assessed using a ranked scale. For example, respondents could be asked to respond to the question 'How important do you think it is to conserve endangered species?' by indicating one of the following: very important; important; unimportant; very unimportant. There is some debate about how many possible responses there should be to such questions. Using an even number of responses (e.g. strongly approve of, approve of, disapprove of, strongly disapprove of) forces the respondent to fall on one side of the argument or the other. There is a danger that those people who feel neutral about the topic may either give an inaccurate indication of their feelings on the matter or will simply ignore the question. Adding a central response which covers the middle ground (i.e. using an odd number of responses) allows respondents to sit on the fence. The final decision depends upon the aims of the survey. The use of three-point scales may miss extreme views (approve, neutral, disapprove), whereas the use of a large number of responses may be confusing (very strongly approve, strongly approve, approve, mildly approve, neutral, mildly disapprove, etc.). It is probably best to choose five or seven responses. An alternative approach is to express the extreme situations (strongly agree and strongly disagree) and present a series of numbers between them, allowing the respondent to select the most appropriate. For example, if we asked the respondents to indicate how strongly they agreed or disagreed with a statement such as 'genetically modified foods are safe to eat', we could ask them to indicate their answer on the following scale:

strongly agree   1   2   3   4   5   6   7   strongly disagree

Data records

Recording data systematically is essential to ensure that everything has been noted and that data do not get lost. It is useful to have recording sheets ready prepared (especially for fieldwork where conditions may be inclement), with as much as possible completed in advance. It is a good idea to use a logbook in which laboratory and field data can be stored, and to copy this out more neatly into a hard-bound book for extra security. Eventually most data will be stored within a spreadsheet on a computer, prior to creating graphs and carrying out statistical analysis. Care needs to be taken at each stage of transcription to avoid errors. Check your data to see whether any obvious mistakes have been made (e.g. double entry, decimal point in the wrong place). Where possible it is helpful to keep the initial paper records in the same format as the final spreadsheet. Such similarity makes it easier to copy from one to the other; copying from a row to a column is especially difficult and prone to error. When organising the final spreadsheet, consider which variables will be recorded in which columns and what each row will signify. This last point is very important when it comes to analysing the data using a computer program (see Appendix B and the relevant chapter for each statistical test). In general, a row is a statistical individual. For example, in a questionnaire survey, you should use a separate column for the answer to each question, or part of a question, and each row would indicate the replies of a separate respondent (so the total number of rows equals the total number of respondents). In a survey of urban air quality using diffusion tubes, the rows should usually be for each diffusion tube while the columns contain details of the site and the contents of the diffusion tube (so the total number of rows equals the total number of diffusion tubes).

It is often useful to begin each recording sheet with a sequentially numbered column indicating the row number (e.g. questionnaire number, quadrat number) and use the same numbering system throughout any logbooks and computer spreadsheets. This speeds up data checking and editing at a later stage. An example of a data recording sheet for use in the field whose format would translate directly to a computer spreadsheet is given in Figure 1.6. This example is for a survey of three rivers. The variables are separated into two types: fixed and measured. Fixed variables are those which are determined during the experimental or survey design: in this example, these are the dates over which the data will be gathered and the codes for each river used (e.g. river X). Measured variables are those obtained from each date and river: pH of the water; flow rate of the water; depth of sediment; presence or absence of a footpath along the river; height of bankside vegetation; percentage cover of vegetation. For clarity in the spreadsheet, it is usual practice to place fixed variables first (e.g. those in columns A and B), to be followed by those which are being measured. New columns, calculated from the existing columns could be incorporated later (e.g. the relative abundance of plant species A, given by the values in column I divided by the values in column H). These days you should not have to calculate new columns by hand,

River survey											
Comments and notes											
Data records	Variables ➤										
	Fixed variables		Measured variables ➤								
	A date of survey	B river sampled (X, Y or Z)	C pH of water	D flow rate of water	E depth of sediment	F presence /absence of a footpath	G height of emergent vegetation	H total % cover of emergent vegetation	I % cover of plant species A	J % cover of plant species B	Etc.
1											
2											
3											
Etc.											

Figure 1.6 *Sample data recording sheet for a river survey*

since computers perform these calculations very easily. In ordering data like this we are making some assumptions about how we will do the analysis. For example, we could see if the pH of the water (a measured variable) differed in different rivers (a fixed variable). Alternatively, we could see if our measured variables are related to each other, separately for each of the fixed variables (e.g. whether the height of emergent vegetation is related to the depth of sediment in each river). Although fixed variables are more often used as factors with which to separate data, sometimes a measured variable may also be used to split a data set. For example, the data could be split on the basis of whether or not a footpath was present (column F).

Summary

By now you should:

● know the general principles of designing experiments and surveys, including the difference between stratified and random sampling strategies, and their applications;

● know how and when to use questionnaires (to sample large numbers of respondents and obtain quantitative data), or semi-structured interviews and unstructured interviews (to obtain an in-depth perspective from a relatively small number of people);

● recognise when data are independent (i.e. when the inclusion of one individual in a sample does not affect the likelihood of any other individual being selected);

● be aware of sources of error, and where possible be able to correct or minimise them through careful experimental and survey design;

- know the difference between nominal, ordinal and interval/ratio, continuous and discrete, derived and non-derived data types;

- be able to record data correctly and organise them in a format suitable for subsequent analysis by computer.

Questions

1.1 Identify the data types represented by these variables:

Variable	Nominal	Ordinal	Interval/ratio
Ratio of a tree's breadth to its height			
Level of trampling on a footpath (low, medium or high)			
Of the invertebrates caught in a trap, the percentage that are herbivorous (plant eating)			
Number of lorries passing a census point within one hour			
Footpath construction (concrete, gravel, none)			

1.2 Identify whether the variables below are continuous or discrete:

Variable	Continuous	Discrete
Amount of rainfall in a day (mm)		
Number of people with an interest in environmental issues		
Percentage of leaves on a tree displaying fungal infection		
The concentration of cadmium in roadside soils		
Type of site (either reclaimed or not)		

1.3 Identify which of the variables below are derived:

Variable	Derived	Not derived
pH of soil		
Number of households that own a car		
Percentage of tree canopy cover in a woodland		
Estimate of tree canopy cover in a woodland (open, intermediate, closed)		
Type of soil (designated as clay, loam or sandy		

1.4 An environmental researcher wishes to measure the plant diversity in an area of 100 × 100 m in a meadow. She decides to take 100 readings using a 1 m quadrat, and considers the following three survey designs. Identify which of the designs is random, systematic or stratified-random, and state briefly any main advantages and disadvantages of each method.

(i) Dividing the sample area into ten 10 m blocks and taking one sample from a random coordinate within each of the blocks.
(ii) Taking random coordinates from the whole 100 × 100 m sample area.
(iii) Setting up ten parallel 100 m lines of quadrats (10 m apart) across the meadow with the quadrat 10 m apart on each of the lines.

1.5 A researcher wishes to know whether there is a difference in the amount of carbon monoxide emitted from the exhausts of large and small cars. He wishes to obtain 20 readings for the carbon monoxide concentrations from each size of car. Should he take 20 readings from a single large car and 20 from a single small car, or should he take a single reading from each of 20 large cars, and from each of 20 small cars? Briefly explain why.

② Describing data

Describing data is an important step in the analysis process as well as in communicating results. This chapter covers:

- **Frequency distributions**
- **Measurements of central tendency and variation in samples**
- **Methods of presenting data in tables and graphs**

Imagine a study into urban environments, where we wish to know the levels of sulphur dioxide in the rainfall. When estimating sulphur dioxide levels, we obtain a sample consisting of several rainfall collections in order to take variation in sulphur dioxide level into account. However, this produces a cumbersome list of numbers to describe the variable. For example, the sulphur dioxide levels (measured as milligrams of sulphur per litre of rainwater) could be:

0.7 0.9 0.9 0.7 0.8 0.7 1.0 0.8 0.8 0.9 0.7 1.2 0.6 0.5 0.8 1.0 1.0 0.8 0.6 0.8

Descriptive statistics are techniques which enable us to describe a list of numbers like this in relatively simple terms. A **statistic** is a measurement based on a sample which (we hope) approximates to the equivalent value (called a **parameter**) from a population. Remember that the term 'population' in statistics means the collection of items from which we sample, in this case potential collections of rainfall in a given period in a given city.

Descriptive statistics

We will first look at tabular and graphical displays that retain all of the information collected, and then examine techniques which condense the data into fewer descriptive terms.

Frequency tables

To more easily examine a set of data such as the one listed above, we could set it out in a frequency table (see Table 2.1), where x indicates each data value (e.g. 0.5 mg of sulphur per litre of rainwater is the first data value, 0.6 mg the second, etc.) and f is the frequency of each value (so that a value of 0.5 mg occurred once, 0.6 mg twice, etc.). Note that the sum of the frequencies, $\sum f$ ($\sum$ is the mathematical symbol instructing you to add all the values together, in this case all of the values in the f column of Table 2.1) should always equal the total number of data points. The number of readings taken is also known as the sample size and is given the symbol n. Here, since 20 readings were taken, $\sum f = n = 20$.

Table 2.1 Frequency table of sulphur dioxide levels in rainfall (mg of sulphur per litre of rainwater)

x	f
0.5	1
0.6	2
0.7	4
0.8	6
0.9	3
1.0	3
1.1	0
1.2	1

Where the data are spread across a large number of values, each with low frequencies, it may be more effective to display the data by combining them into classes of equal sizes. For example, if the rainwater had been analysed using more accurate equipment, the following data (in milligrams) could have been obtained:

0.66 0.89 0.91 0.71 0.80 0.72 0.99 0.78 0.83
0.92 0.72 1.19 0.63 0.49 0.83 0.98 0.96 0.84
0.56 0.78

These data are summarised in Table 2.2. Note that the classes are mutually exclusive, i.e. the boundaries do not overlap, so that each value has only one class to which it can be allocated.

Table 2.2 Frequency table of sulphur dioxide levels in rainfall (mg of sulphur per litre of rainwater) in classes

x	f
0.40–0.49	1
0.50–0.59	1
0.60–0.69	2
0.70–0.79	5
0.80–0.89	5
0.90–0.99	5
1.00–1.09	0
1.10–1.19	1

Frequency histograms

As a more visual alternative to a frequency table, the same information can be displayed on a frequency **histogram** such as that in Figure 2.1, where the size of each block is proportional to the size of the data being represented. This makes it relatively easy to see where the maximum number of data points lie (called the mode) and to see how symmetrical the data are. The shape of the data when plotted on a frequency histogram is called the **frequency distribution**.

Plotting data such as these often results in a pattern where most values lie in the centre, with

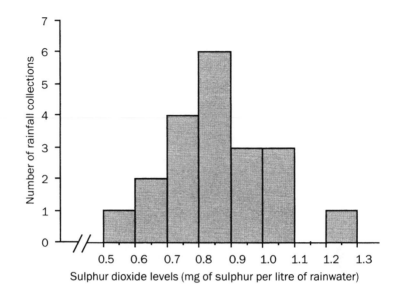

Figure 2.1 *Frequency distribution of sulphur dioxide levels in rainfall (n = 20)*

fewer appearing the further out one goes towards either extreme. With many commonly measured variables (such as mass or height) the more points there are (i.e. the more data has been collected) then the more symmetrical the frequency graph tends to be. For example, a frequency histogram of the levels of sulphur dioxide from 160 rainfall collections (Figure 2.2) is likely to be more symmetrical than that for 20 collections (Figure 2.1).

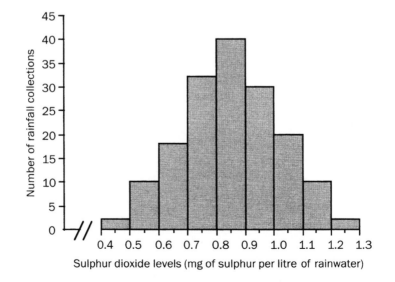

Figure 2.2 *Frequency distribution of sulphur dioxide levels in rainfall (n = 160)*

When large samples of measurement data such as these are collected, the data often conform to a type of distribution called a **normal distribution**. Imagine if thousands of rainfall collections had been made: the resulting histogram could be smoothed to look like that in Figure 2.3, which is close to a normal distribution. The normal distribution is a symmetrical bell-shaped curve defined by a specific mathematical formula. Although we do not need to know the formula, the normal curve is an important concept in statistics because it is an assumption of many statistical tests that the data being analysed are normally distributed (see Chapter 3). Frequency histograms from sampled data can be compared to the ideal normal distribution (see Chapter 3).

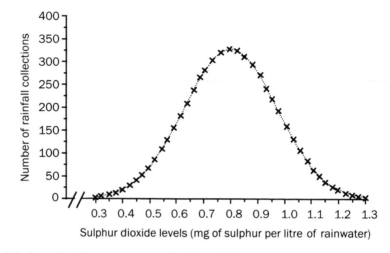

Sulphur dioxide levels (mg of sulphur per litre of rainwater)

Figure 2.3 *Smoothed frequency curve of sulphur dioxide levels in rainfall (n = 7000)*

Data which are not symmetrical are said to be **skewed**. If the heights of trees in a young woodland (say, one only planted 40 years ago) were measured, we might find that there was a cut-off in tree height because although there are many young (short) trees, there are no really old (tall) trees. This situation might create a distribution (shown in Figure 2.4) where there are very few trees over 17.5 m tall. Under these circumstances we need to be careful about the statistical analyses used, or we should manipulate the data so that they do follow a normal distribution. Such manipulations are termed **transformations** and are dealt with in Chapter 3.

Distributions may also be skewed when the data are counts of items (rather than measurements). For example, if the number of plants of a particular species was recorded from each of several quadrats placed randomly in a field, then for a relatively uncommon, randomly distributed species we would expect many of the quadrats to contain no plants. Fewer would contain one plant of that species and fewer still two or more. The type of frequency distribution we would get from this sort of count data is shown in Figure 2.5. Note that the data are displayed as a **bar chart** (with gaps between the bars), and not as a histogram: this is because the

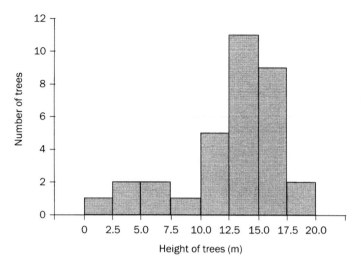

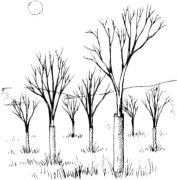

Figure 2.4 *Frequency distribution of the height of trees in a young woodland (n = 34)*

number of plants, on the horizontal axis, is discrete rather than continuous. Where normally distributed data are a requirement of a particular statistical procedure, data following distributions such as that in Figure 2.5 can be transformed (see Chapter 3). Where counts are taken of common items, the data are more likely to be normally distributed and can be treated accordingly.

It is possible to plot data and find that there are two peaks in the frequency histogram (i.e. the distribution is said to be bimodal). In Figure 2.6, which shows the body lengths of toads killed on a major road, two peaks are apparent. In this case we might suspect that we have both adult and juvenile toads, and would have to separate out the adults and juveniles and analyse each separately.

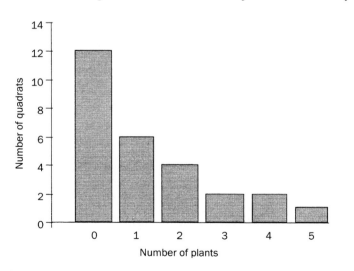

Figure 2.5 *Frequency distribution of number of plants per quadrat (n = 27)*

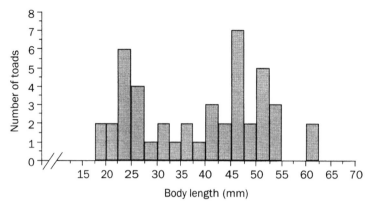

Figure 2.6 *Frequency distribution of toad lengths (n = 45)*

Measures of central tendency

The types of tables and graphs we have looked at so far show patterns in the data (i.e. the distribution). However, they are not always ideal summaries of what is happening: imagine using these techniques to describe the sulphur dioxide levels in rainfall to a colleague over the telephone. Instead, we can use a single value describing the central tendency of the data. There are three commonly used measures.

The **mode** is the value that occurs most frequently in the data; this is most often employed when examining the frequencies of nominal data. For example, in a survey on the use of transport, we might ask 50 people how they travel to work. If 39 go by car, 8 by bus and 3 by train, then the modal value is 'car'.

The **mean** (arithmetic mean) is the most often used average value, calculated by adding all the data points together and then dividing by the number of data points (see Box 2.1). Where the mean has been calculated from the entire population, it is given the symbol μ. Usually we do not know the true value of the population mean μ, and instead calculate a mean from a sample and use it to estimate the population mean. Sample means are given the symbol $\bar{x}$ (pronounced 'ex bar'), and are usually recorded at one decimal place more than the original data. The sample mean is a good estimate of the population mean only if the distribution is symmetrical (i.e. normal).

The **median** is the middle data point when all the data points are listed in numerical order (see Box 2.2). Calculating the median is sometimes useful in situations where a few data points are very much smaller or larger than the majority (known as outliers), since the value of the median is not affected by the magnitude of the end-points, merely by their positions relative to one another.

The median is an ordinal statistic (that is, we need only to know the ranks of the data to calculate it). Suppose we have a situation where we are measuring heights of trees in a wood and the maximum value we can accurately measure is 30 m. We might get the following data (arranged in order of magnitude):

17, 20, 22, 24, 25, <u><u>26</u></u>, 28, 30, >30, >30, >30

We cannot calculate a mean for these data because the final three values are not known to any degree of accuracy. However, we can obtain a median value of 26 m (double underlined).

Box 2.1 *Calculating the mean ($\bar{x}$)*

The mean is usually calculated as follows:

$$\bar{x} = \frac{\sum x}{n}$$

where:

x is a data point;

$\sum x$ is the sum of all of the data points;

n is the sample size (number of data points).

From frequency tables (as in Table 2.1) the mean can also be calculated as:

$$\frac{\sum fx}{\sum f}$$

where:

$\sum fx$ is the sum of all of the data points (calculated by multiplying each value of x by its frequency of occurrence and then summing these);

$\sum f$ is the sample size (calculated as the sum of the frequencies).

The latter formula is only really useful where the data have not been accumulated into size classes (i.e. as in Table 2.1 rather than Table 2.2). If classes of data are used, the x values must be estimated by using the midpoint of each class in the formula above. Because this does not use the actual x values, $\bar{x}$ is less accurate.

In a situation where we get a symmetrical bell-shaped (i.e. normal) distribution, then the central point represents the mean, median and mode. If the frequency distribution is not symmetrical then the values of the mean, median and mode will not be the same; where they differ greatly from each other the data are not normally distributed, and the mean is not an appropriate measure of central tendency.

The mean, median and mode all have the same units as the original data and should be recorded with their units of measurement in a report. It is important to record the number of data points (n) since the larger the sample, the more accurate the measure of central tendency is likely to be as an estimate of the central point of the population.

Having condensed our data into a single measure, we have lost information regarding the shape and range of our data set (i.e. how much variation there is in our data). This

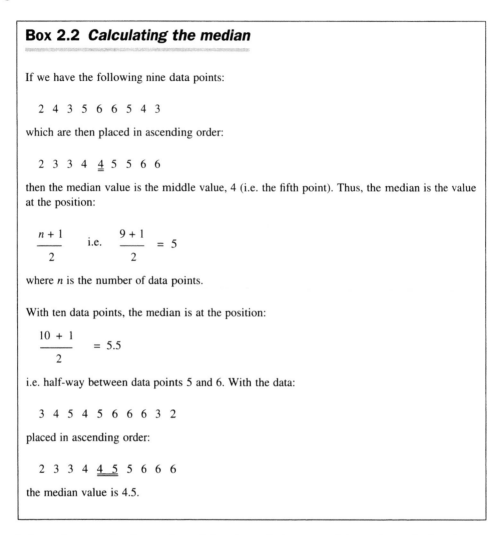

Box 2.2 *Calculating the median*

If we have the following nine data points:

2 4 3 5 6 6 5 4 3

which are then placed in ascending order:

2 3 3 4 4̲ 5 5 6 6

then the median value is the middle value, 4 (i.e. the fifth point). Thus, the median is the value at the position:

$$\frac{n+1}{2} \quad \text{i.e.} \quad \frac{9+1}{2} = 5$$

where n is the number of data points.

With ten data points, the median is at the position:

$$\frac{10+1}{2} = 5.5$$

i.e. half-way between data points 5 and 6. With the data:

3 4 5 4 5 6 6 6 3 2

placed in ascending order:

2 3 3 4 4̲ 5̲ 5 6 6 6

the median value is 4.5.

information can also be condensed, but the techniques used depend on whether the data are approximately normally distributed or not.

Measuring variation of normally distributed data

Measures of central tendency describe one aspect of the data (i.e. what is happening at the middle of the sample). However, this does not provide information about the variation in the data. The difference between the largest and smallest data points (called the **range**) gives us some idea of the variation. However, it can be unduly influenced by extreme values. Suppose we record the number of people per hour using a bottle bank, and arrange the data in ascending order:

11 11 12 12 12 13 13 14 14 14 14 15 15 16 16 22

From these data we can see that a mean of 14 people per hour use the bottle bank. There is a reasonable amount of variation, which we could describe using the range of 22 – 11 = 11 people per hour. However, we can see that this range is heavily influenced by the extreme value of 22, which masks the fact that most of the data points actually lie between 11 and 16 people per hour (a difference of 5, around half the value of the full range). Ideally we could do with a measure which ignores the extreme values and concentrates on the majority of values lying somewhere in the middle of the data set. Where data are normally distributed, we can obtain just such a measure, called the **standard deviation**, the formula for which is given in Box 2.3.

Box 2.3 *Calculating the standard deviation (s)*

The formula for the standard deviation is:

$$s = \sqrt{\frac{\sum (x - \bar{x})^2}{n - 1}}$$

where:

x is each data point;

$\bar{x}$ is the mean of x;

n is the sample size.

1 First, the deviation of each data point from the mean value is calculated $(x - \bar{x})$. This process allocates larger values to those items which are further away from the mean.
2 Next, each value is squared $((x - \bar{x})^2)$, which makes all the values positive and gives an even greater weighting to larger deviations from the mean.
3 The squared values are now summed $(\sum (x - \bar{x})^2)$. This is called the sum of squares of the deviations from the mean – often abbreviated to sum of squares. If you had measured the entire population, you would then take into account the number of measurements by dividing by n. This would obtain the population variance, given the symbol σ^2. However, usually you will have measured a sample from the population, and instead need to calculate the variance (s^2) of your sample. Therefore, divide your sum of squares of the deviations from the mean by $n - 1$ (this $n - 1$ value is known as the degrees of freedom, and will be explained further in Chapter 3, Box 3.1). On pocket calculators, the symbols σ_{n-1} and σ_n are often used to distinguish sample and population standard deviations.
4 The standard deviation (s from a sample or σ from a population) is now calculated by taking the square root of the variance so as to convert the value back into the original units (i.e. if we were looking at height measurements in millimetres, the initial squaring process converted the deviations from the mean into square millimetres; taking the square root at the end converts back to millimetres). The variance and standard deviation are usually recorded to one more significant figure than is the mean.

The following rearrangement of the formula is easier to calculate by hand (see Worked Example 2.1):

$$s = \sqrt{\frac{\sum x^2 - \frac{(\sum x)^2}{n}}{n - 1}}$$

WORKED EXAMPLE 2.1 *Calculating the standard deviation (s) for sulphur dioxide levels in rainfall*

x (mg)	x^2	Calculation
0.7	0.49	Calculate the mean ($\sum x$ is the sum of the x values, and is given at the base of the x column; n is the sample size of 20):
0.9	0.81	
0.9	0.81	
0.7	0.49	$$\bar{x} = \frac{\sum x}{n} = \frac{16.2}{20} = 0.81 \text{ mg}$$
0.8	0.64	Calculate the square of each of the values of x (shown in the second column). From the two columns, use $\sum x$ and $\sum x^2$ and the sample size in the following formula. Notice the distinction between $\sum x^2$ and $(\sum x)^2$: the former is the sum of the squared values (those in the second column) and the latter is the sum of the x values (those in the first column) which is then squared.
0.7	0.49	
1.0	1.00	
0.8	0.64	
0.8	0.64	
0.9	0.81	
0.7	0.49	$$s = \sqrt{\frac{\sum x^2 - \frac{(\sum x)^2}{n}}{n-1}} = \sqrt{\frac{13.64 - \frac{16.2^2}{20}}{19}} = \sqrt{\frac{0.518}{19}}$$
1.2	1.44	
0.6	0.36	$$= \sqrt{0.027\,263\,1} = 0.165\,115\,414 \text{ mg}$$
0.5	0.25	
0.8	0.64	Note that we report the standard deviation to one decimal place more than the mean, so in this example $s = 0.165$.
1.0	1.00	
1.0	1.00	Remember that s^2 is the variance. In this example:
0.8	0.64	$$s = \sqrt{0.027\,263\,1}$$
0.6	0.36	and:
0.8	0.64	
$\sum x = 16.2$ $\sum x^2 = 13.64$		$$s^2 = 0.027\,263\,1$$

The standard deviation gives a measure of how variable the data are. Larger standard deviations (relative to the size of the mean) indicate a larger variation in the data. The calculation of the standard deviation for the levels of sulphur dioxide in rainfall is shown in Worked Example 2.1. For the sulphur dioxide levels, the mean is 0.81 mg of sulphur per litre of rain water and the standard deviation is 0.165 mg. The mean plus one standard deviation (0.81 + 0.165 = 0.975 mg) and the mean minus one standard deviation (0.81 – 0.165 = 0.645 mg) are illustrated on the graph of a normal distribution in Figure 2.7. Because of the mathematical properties of the normal distribution, we know that the area bounded by lines drawn at the mean ± the standard deviation includes 68.27% of the data points. This means that if we randomly drew data points from our distribution, each point would have a 68.27% probability of being in the range $\bar{x} \pm s$. The **probability**, or chance, of an event occurring is an important concept in statistics. A simple description of probability is given in Box 2.4 and probability is discussed further in Chapter 3.

Another important feature of the normal curve is that its exact shape depends on only two values: the mean and the standard deviation. For every different value of the mean and standard deviation there is a unique curve (see Figure 2.8).

Whatever the shape of the normal curve, 68.27% of the data points lie within the range $\bar{x} \pm s$. This concept can be extended to calculate any range around the mean. For example, the range $\bar{x} \pm 2s$ (i.e. lines drawn at 2 standard deviations from the mean) contains 95.44% of the data points. By convention, we are usually interested in the range around the mean which excludes the extreme values and in which the majority of data points lie (i.e. the middle 95% of data points). We can find this range by multiplying the standard deviation by 1.96 (i.e. 95% of the data lie in the range

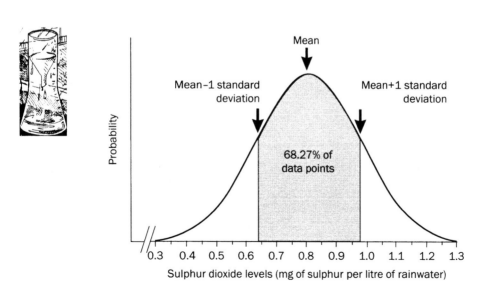

Figure 2.7 *Normal distribution curve for sulphur dioxide levels in rainfall, illustrating the mean and standard deviation*

Box 2.4 Probabilities

Probabilities can be measured either as a percentage ranging from 0% to 100% or as a fraction or decimal between 0 and 1. For example, the probability of obtaining a head when tossing a coin is one in two or 50%, which can also be expressed as 0.5. Similarly, the probability of 68% can be expressed as 0.68 (by dividing by 100). If we predict that an event will occur with a probability of 68%, then out of 100 occasions we would expect the event to occur 68 times. The closer to 1 (or 100%), the more likely the event is to happen. Conversely, the closer to 0 (or 0%) the probability is, the less likely the event is to happen. In statistics, it is often considered that an event occurring with a probability of less than 5% (0.05) is statistically unlikely to happen.

Probabilities may be combined. The probability of tossing a coin and obtaining either a head or a tail is 1.0 (0.5 for the head plus 0.5 for the tail). On the other hand, the probability of tossing a coin twice and obtaining a head both times is 0.25 (0.5 for the first head multiplied by 0.5 for the second). Thus, probabilities are *added* together when we obtain the probability of one event *or* another taking place. They are *multiplied* together when obtaining the probabilities of one event *and* another.

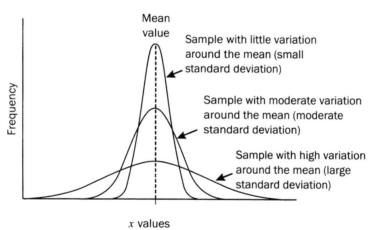

Figure 2.8 *Variability of normally distributed data around the mean*

$\bar{x} \pm 1.96s$). This value of 1.96 is taken from a probability table of z values (see Table 2.3). The table actually gives the probabilities for values lying outside the range, so to find the values where 95% of the data lie inside, we consult the table for a probability (P) of 5% (which is the same as $P = 0.05$). The range in which 95% of the data points occur for the example of sulphur dioxide levels in rainfall is illustrated in Figure 2.9. If we selected data points at random from this population, we would be highly likely (with a probability greater than 95% or 0.95) to select them from the shaded part of the graph (i.e. if we selected 100 points at random, we would expect 95 of them to be in the shaded part of the graph). It would be unlikely (with a

Table 2.3 *Selected values of z for some probability levels (P). Shading indicates the critical values for the example referred to in the text*

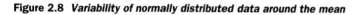

P	0.1	**0.05**	0.01	0.001
z	1.645	1.960	2.576	3.291

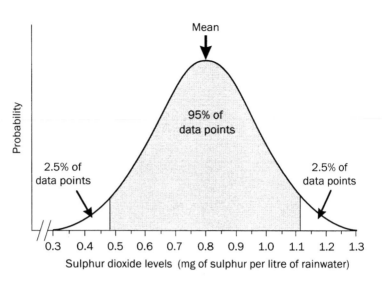

Figure 2.9 *Normal distribution curve for sulphur dioxide in rainfall, illustrating the 95% confidence limits*

probability less than 5% or 0.05) that our randomly selected data point would occur in the unshaded tails of the graph (there is a probability less than 2.5% or 0.025 of selecting a value from the left-hand tail, and likewise from the right-hand tail). If we found a rainwater collection with a sulphur content of 1.2 mg per litre (a value occurring in the right-hand tail), then we could say that it was statistically unlikely (with a probability of less than 0.05) to have come from the population that we sampled to obtain the curve. This 5% value is generally accepted amongst statisticians to represent a relatively unlikely event occurring. This value will become useful in later statistical tests (see Chapter 3).

Reliability of the sample mean

The mathematical properties of the normal distribution also make it possible to calculate how reliable the sample mean is as an estimate of the population mean. The theory behind this is called the central limit theorem. If we took several samples from our population, the mean could be calculated for each one. If these sample means were plotted as a frequency histogram, we would obtain a normal distribution of sample means around an overall mean value of these means. We could then calculate the variation of this distribution of sample means and obtain the standard deviation of the mean of means; this measure is more commonly called the **standard error** of the mean (*SE*). In practice, we do not need to take several samples; instead, using our single sample, the standard error is calculated either by dividing the sample **variance** (obtained in step 3 of Box 2.3) by the number of data points (*n*) and taking the square root of the answer, or by dividing the standard deviation of the sample (obtained in step 4 of Box 2.3) by the square root of the number of data points (see Box 2.5).

Box 2.5 *Calculating the standard error of the mean (SE)*

The standard error can be calculated using:

$$SE = \sqrt{\frac{s^2}{n}} \quad \text{or} \quad SE = \frac{s}{\sqrt{n}}$$

where:

s is the standard deviation;

s^2 is the variance;

n is the sample size.

For example, in Worked Example 2.1, s was 0.165 mg and n was 20, so the standard error is:

$$SE = \frac{0.165}{\sqrt{20}} = 0.037$$

The mean (from Worked Example 2.1) and standard error can be displayed as:
0.81 ± 0.037 mg of sulphur per litre of rainwater.

Following the logic of the central limit theorem, another measurement of the reliability of the sample mean is the range within which 95% of possible sample means lie. We call this measurement the 95% **confidence limits** of the mean and, if the number of data points is large (usually 30 or more), then we simply multiply the standard error by the value of 1.96 (from the z table for $P = 0.05$: Table 2.3). However, if the sample size is small (n is less than 30) then we are less confident that our sample standard deviation is a reliable estimate of the standard deviation of the population, and we need to use a correction factor to allow for this. Instead of z values, we use a table of t values (an extract of which is shown in Table 2.4) where the value required is dependent not only upon the level of confidence required (here 95%), but also on the number of data points upon which the calculation of the mean is based. The table is

Table 2.4 Selected values of t for some probability levels (P). Shading indicates the critical values for the example referred to in the text. A more comprehensive table of t values is given in Table D.2 (Appendix D)

df	t	
$(n-1)$	$P = 0.05$	$P = 0.01$
16	2.120	2.921
17	2.110	2.898
18	2.101	2.878
19	2.093	2.861
20	2.086	2.845
⋮	⋮	⋮
∞ (infinity)	1.960	2.576

Box 2.6   *Calculating the 95% confidence limits of the mean*

For large samples ($n \geq 30$):

  95% confidence limits = 1.96 SE

For small samples ($n < 30$):

  95% confidence limits = $t_{0.05\ [n-1]}\ SE$

Where the subscript of the t symbol, 0.05 [n – 1], instructs you to look up t at the 0.05 level (highlighted in bold in Table 2.4) with degrees of freedom of n – 1, i.e. the sample size (n) minus 1.

To calculate the 95% confidence limits, look down the $P = 0.05$ column in Table 2.4 and select the t value at the appropriate degrees of freedom. Then multiply the standard error by the t value. From Worked Example 2.1, the mean was 0.81 mg of sulphur per litre of rainwater, derived from 20 data points. From Box 2.5 we found that the standard error of the mean was 0.037 mg. The t value at $P = 0.05$ and 19 degrees of freedom (following the shaded parts of Table 2.4) is 2.093. Multiply this t value by the standard error (± 0.037 mg) to give the 95% confidence limits for the mean as ± 0.077 mg. The mean and the 95% confidence limits can be displayed as: 0.81 ± 0.077 mg of sulphur per litre of rainwater.

---

consulted using the sample size minus 1 (see the shaded area of Table 2.4). This corrected sample size, called the **degrees of freedom** (df), takes into account the fact that we are making estimates from a population (degrees of freedom are explained further in the next chapter: Box 3.1). From Table 2.4 it can be seen that as the sample size increases, the t value becomes closer to the equivalent value of z (at an infinitely large sample size, df = ∞ and $t = z$). Box 2.6 demonstrates the calculation of the 95% confidence limits of the mean.

Measuring variation of ordinal or non-normal data

For data which are unsuited to the calculation of means and standard deviations (i.e. are not normally distributed, and where the median is the appropriate measure of central tendency) there is a useful measure of variability called the **interquartile range** (see Box 2.7). Here, the median represents the middle data point of a sample which has been sorted in ascending order, the lower **quartile** (Q_1) is found at the point halfway between the median and the lowest value and the upper quartile (Q_3) is halfway between the median and the highest value.

Methods of presenting data

Summaries of the data (but not normally raw data) are presented in the results section of a report, either in the form of graphs, tables or in the text. All data should be

---

Box 2.7 *Calculating the interquartile range*

The interquartile range is simply the difference between the lower quartile (Q_1) below which 25% of the data points lie, and the upper quartile (Q_3) above which again 25% of the data points lie. Where the number of data points does not lend itself to easy identification of the quartiles, the following can be used:

- Q_1 is the nearest value to that indicated by a rank of $(n + 1)/4$
- Q_3 is the nearest value to that indicated by a rank of $3(n + 1)/4$

where n is the number of data points.

For example, the following data estimate (on a scale of 1 to 6) the weathering of stonework on a city building (where 1 represents no obvious weathering and 6 represents over 80% of the face of the stone weathered). In ascending order the data are:

2 3 3 3 <u>3</u> 4 4 4 4 4 <u>4</u> <u>4</u> 5 5 5 <u>5</u> 6 6 6 6

The median position is double underlined and the quartiles are single underlined. The median value here is 4 (i.e. the position between $(n + 1)/2 = (20 + 1)/2 = 11.5$, or between the 11th and 12th position, since both of these values are 4, we calculate $(4 + 4)/2 = 4$). The lower quartile (Q_1 is 3 (calculated by $(20 + 1)/4 = 5.25$, i.e. the fifth value in ascending order) and the upper quartile (Q_3) is 5 (calculated by $3(20 + 1)/4 = 15.75$, i.e. the 16th value in ascending order).

---

displayed with the units of measurement (SI units: see Box 2.8). The presentation methods used depend on the data and the reasons for display. Graphs (which should always be labelled as figures) and tables are useful for different reasons. As a general rule, when deciding how to present your data, illustrate the most interesting or important results as graphs (i.e. the aspects of your report which you want the reader to focus on). Avoid having too many pages of graphs, as this can be hard for the reader to absorb; instead, use tables to summarise several results or to present detailed information without overwhelming the reader. The text should both guide the reader through consecutively labelled figures and tables, and summarise all the results. In other words, the reader should be able to see what you found from the text, and know which table or graph to look at for the details. Both figures and tables should have an appropriate (brief) title and be labelled so that they stand alone (i.e. so the reader can understand them without having to consult the text).

Presenting measures of central tendency

Whether you use text, tables or figures to display your results, always report the appropriate summary statistics: for example, mean values with the correct SI unit, the sample size (n) and one of the measures of variation: standard deviation, standard error or 95% confidence limits (state which one you are using). As a rule of thumb, if you report a single mean and associated measure of variation, then it is usual to use

Box 2.8 *SI units of measurement*

To standardise the units in which measurements are made, an international system has been agreed (Système International d'Unités, abbreviated to SI units). The standard units for major types of measurement are shown below, together with the common prefixes indicating multiples of SI units:

Measurement	SI unit (symbol)	Example of use
Length	metre (m)	1.5 m
Mass	kilogram (kg)	0.25 kg
Time	second (s)	45 s
Temperature	kelvin (K)	273.1 K
Amount of substance	mole (mol)	1 mol

The prefixes help to keep the numbers to manageable levels, usually so that the measurement lies between 0.1 and 1000 (e.g. 1 mg rather than 0.000 001 kg, and 100 km rather than 100 000 m). Under this system deci (d) meaning 10^{-1} and centi (c) meaning 10^{-2} are not supposed to be used. Note that the convention uses spaces (not commas) to separate groups of three digits before and after the decimal point, and a space between the number and the unit. Where units are combined in derived measurements (e.g. speed in metres per second), a space is left between each of the units to avoid misunderstandings. For example, ms is milliseconds and m s^{-1} is metres per second (note that per is denoted by a negative index). It is likely that some non-standard usage will continue for some time, especially where measurements require large prefixes (e.g. time measured in seconds) or where there are more than 1000 units between the stages (e.g. for area 1 m^2 = 1 000 000 mm^2, and for volume 1 m^3 = 1 000 000 000 mm^3). In such cases alternatives are likely to remain commonplace, for example the use of hours or days for time, hectares for area (1 ha = 0.01 km^2 or 10 000 m^2) and litres for volume (1 litre = 1 × 10^6 mm^3 or 1 × 10^{-3} m^3). Temperature is often still measured in degrees Celsius rather than in kelvin, where, although the units are of the same value, the scales have different starting points.

Common prefixes (abbreviation)		
10^{-9}	nano	(n)
10^{-6}	micro	(µ)
10^{-3}	milli	(m)
10^3	kilo	(k)
10^6	mega	(M)
10^9	giga	(G)

the standard deviation. When you compare more than one mean, then the standard errors or confidence limits should be used. When using medians, also report the interquartile range and the sample size. Mean or median values can also be reported in the text if you do not want to include them in a table or a graph. For example, in a survey of toads found as road kills, if you had one mean value to report, you could state in the text that:

the mean (± standard deviation) mass of toads was 14.3 ± 3.39 g, $n = 28$.

If you had several toad measurements, perhaps from more than one age group, you may prefer to summarise them in a table. In Table 2.5, variables measured (mass, length, etc.) are placed along the top of the table, and the age classes (juveniles and

Table 2.5 *Mean (± standard error) measurements of juvenile (n = 17) and adult (n = 28) toads*

	Mass (g)	Length (mm)[a]	Jaw width (mm) etc. →
Juveniles	3.2 (± 0.23)	24.4 (± 1.00)	10.3 (± 0.24)
Adults	14.3 (± 0.64)	46.8 (± 1.30)	17.7 (± 0.40)

[a] Measured as snout–vent length.

adults) are arranged down the side. This arrangement of variables horizontally and of classes vertically is preferable because it is easier to compare the numbers by eye if they are in columns (you can pick out easily that adult toads are larger by scanning down the first column). It is not as easy to compare if the table is arranged the other way around, i.e. with each measured variable being read as a row (although this may be unavoidable if you have too many measured variables to fit in as columns). Note that, to avoid cluttering the table, there is no extra column for sample sizes (they are summarised in the title instead), and that extra details can be presented as footnotes.

Alternatively, you may wish to present your means in graphical form, using a bar chart (Figure 2.10a). Here you would show a measure of the variation by using error bars (state in the title of the graph whether the bars are standard deviations, standard errors or confidence limits), and include the sample size (either at the base of the bar or in the title of the graph). In the bar chart in Figure 2.10a, the error bars are quite small and it may be difficult for a reader to identify the values. You could emphasise the differences and errors by starting the axis for toad length at 20 mm rather than 0 mm. However, so as to not mislead the reader into thinking the differences are bigger than they actually are, you should include a break in the axis and in the bars (and their shading). This is easier to do in a hand-drawn graph than a statistics/ graphics computer package. An effective alternative way to display the means is to use a **point chart**, where you only have to show the break in the axes (see Figure 2.10b).

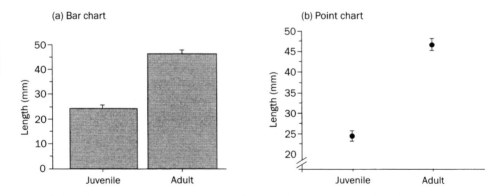

Figure 2.10 *Mean lengths of juvenile (n = 17) and adult (n = 28) toads. The bars represent standard errors*

Medians should be quoted with their interquartile range and sample size. For example, from a survey of trees on polluted and clean sites, where trees were given a condition score of 1 (poor) to 6 (good), you could say that the median condition of trees grown on the polluted sites is 3.5 ($Q_1 = 2$, $Q_3 = 4$, $n = 10$). Alternatively you could display the medians, quartiles and ranges using **box and whisker plots** (Figure 2.11). In such plots, the median is drawn as a line at the appropriate level and is surrounded by a box, the ends of which define the quartiles (Q_1 and Q_3). The whiskers are drawn to the lower and upper limits of the data. Some computer programs (such as the one used to create the box and whisker plot in Figure 2.11) draw the whiskers to the values covering most data points (usually at the 10% and 90% points) and show the minimum and maximum values as separate points. Note that if several elements have the same value (as do the upper quartile, 90% whisker, and maximum value for the trees on the clean sites in Figure 2.11) they will overlap on the graph. If the median has the same value as one of the quartiles, a thicker line may be used for the median to distinguish it from the other quartile.

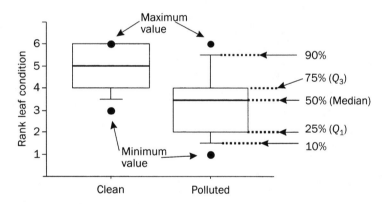

Figure 2.11 *Rank leaf condition of trees found on polluted and clean sites (n = 10 for each). Higher scores indicate leaves in better condition*

Presenting relationships between variables

So far, we have examined distributions where we have measured a variable in one or more discrete categories (for example, measuring the variable toad length in two different categories, adult and juvenile). However, if we had measured the mass of each toad in addition to its length, we could examine the relationship between the two continuous variables, mass and length. We can examine relationships between two measured (or ranked) variables when, for every individual sample unit (e.g. an individual toad), we have two measurements. We would not be very surprised if we plotted toad weight and length and found that, as length increased, so did weight. As a

Figure 2.12 *Relationship between the pH of the soil and the number of plant species growing at 13 colliery spoil sites*

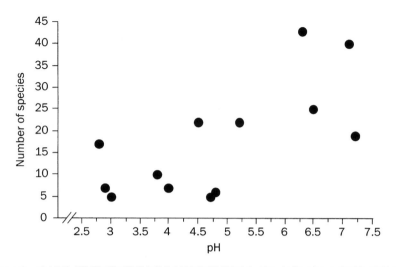

more interesting example, we could examine the relationship between the pH of colliery spoil and plant colonisation. We could survey several colliery sites that were as similar as possible in terms of size, time since the spoil was dumped, etc. For each site (which is a sample unit) two variables would be measured: the number of plant species in a given area and the pH. From this we might find the relationship shown in the **scatterplot** in Figure 2.12. Relationships between variables are discussed in Chapter 5. Note that it is not necessary (and in some cases it is incorrect) to draw a line on a graph such as this. If we wanted to fit a line (and were justified in doing so), then this is done mathematically (see Chapter 5) and never by eye.

Cross-tabulation of data

Cross-tabulation is a useful method for presenting data which consist of frequencies in nominal categories (often used for questionnaire data). For example, in a study of the demographics of people who recycle, we might split items that are most commonly recycled by the sex of the respondent. Separate frequency tables (such as those given

Table 2.6 Frequency tables of items most commonly recycled and sex of respondent

(a) Items most commonly recycled

Item	Frequency
Newspapers	22
Glass bottles	25
Plastic containers	23
Tins	10
Total	80

(b) Sex of people recycling items

Sex	Frequency
Male	30
Female	50
Total	80

Table 2.7 Cross-tabulation of items most commonly recycled and sex of respondent

Commonest item recycled	Sex		Total number of people recycling
	Males	Females	
Newspapers	number = 12 row % = 54.5 column % = 40 overall % = 15	number = 10 row % = 45.5 column % = 20 overall % = 12.5	22 people
Glass bottles	number = 10 row % = 40 column % = 33.33 overall % = 12.5	number = 15 row % = 60 column % = 30 overall % = 18.75	25 people
Plastic containers	number = 5 row % = 22 column % = 16.67 overall % = 6.25	number = 18 row % = 78 column % = 36 overall % = 22.5	23 people
Tins	number = 3 row % = 30 column % = 10 overall % = 3.75	number = 7 row % = 70 column % = 14 overall % = 8.75	10 people
Total number of each sex	30 males	50 females	80 people

in Table 2.6) could be used to find the frequency of sexes and the frequency of the most commonly recycled item. However, we would need another technique to see whether one sex recycled one type of item more than the other. Cross-tabulation generates tables with rows to represent one variable and columns to represent the other, the cells of the table indicating the frequency of responses. For example, in Table 2.7 the column variable is sex (males, females) and the row variable is the item which is recycled the most on each visit. In the first cell is the number of visitors who are male and whose major item for recycling is newspapers (12 people). Percentages can be calculated based on the row totals, column totals or the overall number of respondents. Row percentages would indicate how many of those mainly recycling a particular item were male or female (e.g. 54.5% of respondents mainly recycling newspapers were male). Column percentages would give the percentage of each sex mainly recycling particular items (e.g. 40% of males mainly recycled newspapers), while overall percentages would be the percentage that a particular sex mainly recycling a particular item was of the total number of respondents (e.g. 15% of respondents were males who mainly recycled newspapers). Frequency data such as these can be further analysed using the techniques described in Chapter 6.

Cross-tabulation tables may also be used to report the results of another measured (or ranked) variable which is separated by the row and column categories. So, for example, the numbers of each of the commonest items recycled could be recorded for each person and the means (together with the standard errors and the number of data points) could be reported in each of the cells split by sex and type of item (Table 2.8). For example, the 12 males who were mainly recycling newspapers, recycled an average of 11.4 ± 0.79 newspapers. Further analysis of data such as these is carried out using techniques described in Chapter 7.

If graphical representation of the frequency of nominal variables is required, then bar charts are appropriate. Figure 2.13a shows the most frequently recycled materials split into categories. Here, it is easy to see that the modal (most frequently occurring) value is glass bottles. Note that in bar charts there are spaces between the bars, unlike the graphical representation of a continuous variable (e.g. the sulphur dioxide levels in rainfall data from earlier – Figure 2.1) which is a histogram and has no such spaces. An alternative to the bar chart, especially useful when there are many subdivisions, is a **pie chart** or pie diagram. Figure 2.13b shows the recycled item data in this form.

Table 2.8 Cross-tabulation of mean number of the commonest items recycled (± standard error) split by type of item and sex of respondent

Commonest items recycled	Sex		Across all people recycling items
	Males	Females	
Newspapers	11.4 (± 0.79) n = 12	9.6 (± 0.73) n = 10	10.6 (± 0.57) n = 22
Glass bottles	10.5 (± 0.43) n = 10	6.8 (± 0.30) n = 15	8.3 (± 0.44) n = 25
Plastic containers	7.6 (± 0.51) n = 5	6.9 (± 0.30) n = 18	7.1 (± 0.26) n = 23
Tins	9.3 (± 0.88) n = 3	6.4 (± 0.81) n = 7	7.3 (± 0.75) n = 10
Across all items	10.3 (± 0.43) n = 30	7.4 (± 0.28) n = 50	8.4 (± 0.28) n = 80

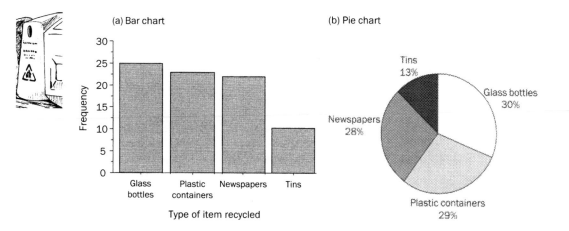

Figure 2.13 *Frequency of items most commonly recycled*

Pie charts are especially suitable where comparisons of the frequencies are being made against the whole; for example, it is easier to see from the pie chart that in nearly a third of the total number of cases glass bottles are the commonest item recycled, but it is more difficult to see that it is the modal value.

General guidelines for drawing graphs and charts

There are several points which can help to ensure clarity in presenting data:

- both axes should be labelled and the units used should be specified;
- thin (rather than thick) outlines and small symbols should be used to allow more precision in interpreting the graph;
- the scales of each axis should be appropriate to the data being used (i.e. not necessarily beginning with zero but remembering to put a break in the axis if not starting at zero) and making full use of the axis lengths available;
- when illustrating frequency distributions of a continuous variable (such as lengths or masses) use a histogram (a histogram has no spaces between the blocks), but when illustrating frequency distributions of a discrete variable (such as counts or nominal categories) use a bar chart (with spaces between the bars).

Summary

By now you should be able to:

- represent frequency data of a continuous variable as a frequency table or histogram;
- visually identify the differences between normally and non-normally distributed (e.g. skewed or bimodal) data from a histogram;

- calculate the three measures of central tendency (i.e. mean for normally distributed data; median for non-normally distributed or ordinal data; and mode for nominal data);

- calculate measures of variation (standard deviation for normally distributed data; interquartile range for non-normally distributed or ordinal data), and, for normally distributed data, be able to calculate the reliability of the estimate of the mean (using standard errors and 95% confidence limits);

- draw appropriate graphs representing the means, medians and measures of variation (bar charts or point charts with error bars for normally distributed data, and box and whisker plots for non-normally distributed or ordinal data);

- produce appropriate output for other types of results (e.g. scatterplots for relationship data, and cross-tabulation, bar charts and pie charts for nominal frequency data).

Questions

2.1 Twenty hedgehogs were sampled from a population in autumn (before hibernation) and weighed. In the spring, the population was again sampled and twenty hedgehogs were captured and weighed. The following data (in kilograms) were obtained:

Autumn

0.91 0.87 0.97 0.79 1.09 0.96 0.98 0.82 0.84 0.89

1.01 0.92 0.70 0.88 0.91 0.89 0.79 0.81 0.99 0.75

Spring

0.71 0.55 0.60 0.74 0.58 0.49 0.61 0.66 0.75 0.85

0.61 0.49 0.59 0.83 0.74 0.77 0.82 0.57 0.66 0.71

(i) Create a frequency table for the mass of hedgehogs caught in autumn.
(ii) Draw a frequency histogram of hedgehog mass in autumn.
(iii) For both spring and autumn hedgehog collections, calculate the statistics indicated (t values – needed for the calculation of the 95% confidence limits of the mean – are given in Table D.2, Appendix D):

	Spring	*Autumn*
Mean		
Standard deviation		
Standard error of the mean		
Degrees of freedom		
95% confidence limits of the mean		

(iv) Plot the means and standard error bars of hedgehog mass in the autumn and spring collections on a bar or point chart.

2.2 In a survey to identify the level of public concern regarding two environmental topics (genetically modified food and intensive pesticide use), 20 people were asked to score their level of concern on a scale from 1 (not concerned at all) to 5 (very concerned). The results obtained were:

Genetically modified food   5 1 1 4 3 3 4 5 2 3 5 4 3 2 1 5 1 2 2 3

Intensive pesticide use   2 1 2 1 1 2 3 1 2 3 3 1 2 1 1 1 1 1 1 2

(i) Calculate the statistics below for the level of public concern regarding each topic:

	Genetically modified food	*Intensive pesticide use*
Median		
Lower quartile		
Upper quartile		

(ii) Illustrate the data using box and whisker plots.

③ Using statistics to answer questions

Inferential statistical methods provide an objective way of deciding whether there are differences between samples, relationships between variables and associations between frequency distributions. This chapter covers:

- **Hypothesis testing**
- **Data transformations**
- **Choosing a statistical test**

Displaying data using means or medians allows us to summarise many points and to examine some of the attributes and patterns of a sample. However, interpretation of such summaries tends to be descriptive and may be superficial. Frequently we wish to draw scientific conclusions as to whether there are:

- differences between categories or treatments (e.g. in the nature conservation value of different sites, in the heavy metal concentration of several types of soil, or in the development of environmental awareness in children of different backgrounds);
- relationships between variables (e.g. between growth rate and nutrition, or pollutant concentration and distance from the pollutant source);
- associations between frequency distributions (e.g. the frequency with which two species of plant are found together, or the proportion of people choosing organic produce over non-organic).

Inferential statistics provides us with objective methods for deciding whether such differences, relationships or associations are significant, i.e. whether they are likely to be real, or whether they are more likely to have arisen because of chance.

Hypothesis testing

In Chapter 2, we looked at the means and standard errors of the lengths of juvenile and adult toads in order to describe the data set. We might wish to take this further to discover whether the apparent difference between two means is a significant one (i.e. whether it is large enough to be useful in the context of the investigation): this is where inferential statistics is used. The first step in the process of performing inferential statistics is to form a **hypothesis**. A statistical hypothesis may be different from the scientific hypothesis of interest: in a statistical test where we compare two means, we actually test the hypothesis that there is no significant difference between the means. Similarly, if we are investigating the relationship between two variables, we test the hypothesis that there is no significant relationship. This 'no difference' (or no relationship) hypothesis is known as the **null hypothesis** and is denoted as H_0. To take an example, imagine that we are investigating whether there is a difference in the radioactivity of the surrounding areas near two (imaginary) types of nuclear power generation plant, one type employing the 'Rayzoom' process and the other the 'Power Gush' process. One measure of radioactivity we could use is the activity of strontium-90 present in milk from the closest farm to each of ten Rayzoom plants and to each of ten Power Gush plants (i.e. 20 farms in total). The null hypothesis is that there is no difference in strontium-90 activity in milk from farms near to the two types of plant. If, after a statistical test, we find that the data do not support the null hypothesis then we would accept an **alternative hypothesis** which must be set up in advance of the

investigation. For some tests (such as those where we compare two means or examine relationships between variables), there are two types of alternative hypothesis: **specific** and non-specific. In a specific test, we specify which direction the differences or relationship will lie, and then only test in that direction. For example, the null hypothesis could be that the mean of sample A is not larger than that of sample B, and the alternative hypothesis could be that the mean of sample A is larger than that of sample B. However, we are usually interested in the case where the difference could be in any direction. In this case, the **non-specific** alternative hypothesis would be that the means of sample A and B are different. In the example of strontium-90 activity in milk, we have no clear reason in advance of the survey to predict the direction of the difference (if any) in the level of radioactivity near to Rayzoom or Power Gush plants. We would, therefore, set up a non-specific alternative hypothesis that there is a difference in strontium-90 activity in milk from farms near to the two different types of nuclear plants.

The next stage, after specifying the null hypothesis, is to calculate a **test statistic**. The test statistic is a single number calculated from the data. In the same way as the distribution of data values often follows a normal distribution, test statistics also have known probability distributions. This means that any given test statistic value has an associated probability value. The interpretation of a statistical test rests upon this probability of the test statistic occurring by chance. A test statistic is often based on the ratio of explained variation to that which is unexplained. The explained variation

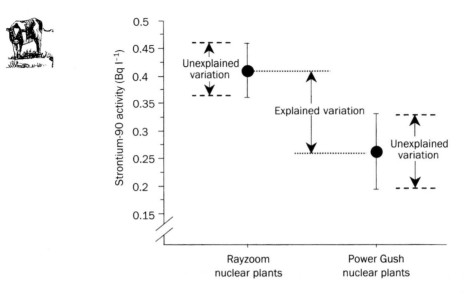

Figure 3.1 *Illustration of the two types of variation in the activity of strontium-90 in milk near to two types of nuclear plant (n = 10 of each type). The bars are standard deviations which represent the unexplained variation, whilst the difference between the means represents the explained variation*

(or variation between samples) is that which can be attributed to the experimental or survey design: e.g. in the strontium-90 activity in milk example, the explained variation would be due to a farm's proximity to a particular type of nuclear plant (Rayzoom or Power Gush). Unexplained variation is the variation of the individual samples (e.g. within each of the two samples of the farms). The greater the ratio of the explained to unexplained variation, then the more likely the means are to be significantly different. The two types of variation can be visualised in Figure 3.1.

Later in this chapter we consider how to select a test, and then the procedure for calculating the test statistic for each particular test is given in subsequent chapters (Chapters 4–7). Since test statistics have known mathematical properties, if we know the size of the samples (often adjusted to take into account the number of fixed points there are – the degrees of freedom – see Box 3.1) we can find the probability of obtaining any particular test statistic value by chance. In other words, we obtain a probability (P) of the null hypothesis being true (in the case of a difference test, the probability of there being no difference). If this probability is high (P greater or equal to 0.05) then there is a high probability of the null hypothesis being true and we therefore accept the null hypothesis that there is no significant difference. However, if the probability is low (P less than 0.05) then there is a low probability of the null hypothesis being true and we can reject the null hypothesis of no difference and accept an alternative hypothesis that there is a difference. If we find a difference, it is known as a statistically significant difference and we report it as such, giving the probability. The method of reporting the probability is given in Box 3.2. The 0.05 (or

5%) level of significance is called the **critical level** and is an arbitrary limit that is accepted by most statisticians. A probability of 0.05 implies that the result would occur purely as a result of random sampling fluctuations 5 times in 100 (or 1 in 20). In other words, if there was no significant difference in the strontium-90 levels from milk near to the two types of nuclear plant, but we repeated the survey 20 times, we would expect to obtain a significant difference once simply by chance. In statistics, we can never be absolutely certain that rejecting the null hypothesis is the correct outcome. The probability of rejecting the null hypothesis when in fact it is true is called a **type I error** (the term 'error' does not imply mistakes on the part of the researcher or statistician: it is an inherent hazard in the process of performing statistical tests). By reducing the significance level we test at, we could be more certain that a type I error would not occur (by testing at the 0.001 level, we could be 99.9% confident that we were not falsely rejecting the null hypothesis). However, there is another sort of statistical error (called a **type II error**) that becomes more likely to occur as the critical level is reduced: that is the probability of accepting the null hypothesis when in fact it is false. For example, we might conclude that there

---

Box 3.1 *Degrees of freedom* (df)

When measuring the degree of scatter of data points around the mean (i.e. the standard deviation), we took sample size into account by dividing by a number called the degrees of freedom which was the sample size minus 1 (see step 3, Box 2.3). We use the degrees of freedom (sometimes given the symbol v) rather than the sample size because we want to account for the number of independent estimates of scatter we have sampled from the population. As soon as we estimate a parameter by calculating a statistic (such as the mean), we lose an independent estimate of the scatter. In the case of the standard deviation, the mean has already been calculated and is then used in the formula for the standard deviation. Now that the mean is known, although we still have a sample size of n, only $n - 1$ of these samples are independent. For example, if a mean of 5.0 was based on three items, and two items were drawn at random and found to be 4.0 and 5.0, the final sample cannot be drawn randomly: it must be 6.0 in order for all three data points to have a mean of 5.0. After calculation of the mean from three independent data points, only two remain independent, and therefore the df = 3 − 1 = 2.

There are not always $n - 1$ degrees of freedom: the number depends on how many population parameters are estimated, which in turn depends on the statistical test being used. The formula for calculating the number of degrees of freedom for each of the tests is given as we encounter the tests in subsequent chapters.

For another suite of tests (called nonparametric tests), degrees of freedom often do not exist. This is because these tests do not use any estimated parameters from the data (hence the name nonparametric). Only the positions of the samples in relation to each other (their location or ranks) are used in the calculation of the statistics (e.g. in the measure of variation, the upper quartile is simply the position below which three quarters of the data points lie, and the lower quartile is the position below which one quarter of the data points lie – see Chapter 2). Thus, when consulting tables of many nonparametric test statistics, we often use the number of data points (n) without any adjustment. Nonparametric tests are described in more detail later in this and the following chapters.

Box 3.2 *Reporting the probability* (P)

After calculating a test statistic, most computer programs display the exact probability of that test statistic arising by chance. However, as we will see in the following chapters, when calculating test statistics by hand, we need to look up the test statistic in a statistical table to obtain the P value. Here, we do not get an exact value for the probability. In the statistical tables given at the end of this book (Appendix D), the test statistic values are given for $P = 0.05$ and $P = 0.01$. Examination of these tables will allow you to record whether P is greater or equal to 0.05 (i.e. $P \geq 0.05$) and therefore is not significant, or whether P is less than 0.05 ($P < 0.05$) or less than 0.01 ($P < 0.01$) and is significant. If your computer gives the exact probability value, then use it when you report the results.

It is common to find that significant P values are highlighted in a results section, using * next to $P < 0.05$, and ** next to P less 0.01. You may find it useful to highlight values in a large table (using a **bold** type face for any significant P values is also effective). It is also common practice in the results text to refer to results where $P < 0.05$ as significant, and $P < 0.01$ as highly significant. In addition to the P value, always report the test statistic and degrees of freedom (or sample size for some tests). Note that P may be called α in some books or computer programs.

---

was no difference between two means, when in reality there was a difference. For most purposes, the 0.05 level provides an acceptable compromise between obtaining type I and type II errors.

There are occasions where the use of a lower critical limit is advisable. One such occasion is where a large number of tests are performed. If 20 tests are computed, at the 0.05 level we would expect at least one result to appear significant by chance. One method to get around this problem (the Bonferroni method) is to reduce the critical level according to the number of tests being performed. The usual critical level (0.05) is simply divided by the number of tests, so that when 20 tests are calculated, the critical level for rejecting the null hypothesis would be $0.05 \div 20 = 0.0025$ and with 50 tests it would be $0.05 \div 50 = 0.001$. For more sophisticated analyses, see Sokal and Rohlf (1995).

If we establish that a result is significant, we need to return to the original data (e.g. by examining a plot of the means or a scatterplot of the relationship) to identify the direction of any difference or relationship. Figure 3.2 summarises the stages in conducting a statistical test when we have a non-specific alternative hypothesis (as is usually the case) and shows the two outcomes depending on whether the test statistic is significant or not. Under certain circumstances, we are justified in having a specific alternative hypothesis. For example, we may wish to examine the impact of an antifeedant (a chemical which, in invertebrates, can reduce the urge to eat) on the size of pests on a crop. We could add the chemical to one set of plants and water to another set. When we compare the sizes of individual pests on the two samples of plants we would expect that if there were any differences, those exposed to the antifeedant to be smaller than those on the control plants. We would then test the null hypothesis that the pests on experimental plants were not significantly smaller than

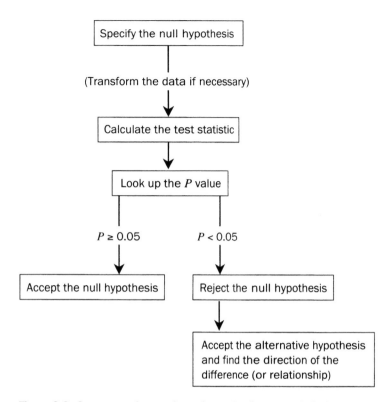

those on control plants (with the alternative hypothesis being that they are significantly smaller). To test this specific hypothesis, we would compute the test statistic as usual and simply halve the resultant probability value. The reason for halving the probability is that we are only interested in a situation where pests on experimental plants are smaller than those on controls (and not the 50% of occasions where random sampling fluctuations would result in pests being larger on experimental plants). However, because it is fairly rare that we predict the direction of the result in advance, many statistical computer programs only give the P value for non-specific tests. It should be noted that you must have sound theoretical or experimental grounds for wishing to use a specific hypothesis: halving the probability later purely as a means of obtaining a significant result is cheating. A specific alternative hypothesis can also be set up for relationship questions. For example, we might hypothesise that the greater the amount of antifeedant used, the smaller the animals living on the plants, i.e. we have a specific alternative hypothesis of a negative relationship.

Figure 3.2 *Sequence of procedures in conducting a statistical test*

Statistical tests

There are many different statistical tests, some of which are specialised and used relatively rarely. Others are very commonly used and, unfortunately, many are often abused (even within the scientific and social scientific press). There are defined circumstances when certain tests are used and how they are then applied and interpreted. You will slowly gain experience in using the common tests, but you should keep in mind that there may be alternatives available for particular circumstances. A table is given later (at the very end of the book) which shows those tests covered in this book. You should follow the further reading suggestions, or seek advice if you feel that the question you are asking requires a special type of analysis.

Most tests fall into one of two types: **parametric** (which are based around normally distributed data) and **nonparametric** (also called distribution-free tests). Although parametric tests are perhaps more commonly used, there are a certain circumstances (given in Table 3.1) where nonparametric tests should be used in preference.

Table 3.1 Attributes of parametric and nonparametric tests

Parametric tests	Nonparametric tests
Require interval/ratio data (used on actual observations or measurements)	May be used on nominal or ordinal data as well as interval/ratio data (can be used on observations, measurements or ranked values)
Require data to be normally distributed (some derived variables, and non-normal and count data must be transformed)	Do not require normally distributed data (transformation of data not necessary)
With small sample sizes it is difficult to check that the data are normal	May be used on relatively small sample sizes since normality is not needed
Use all the information in the data	Do not use all the information in the data (ranks are used as opposed to the actual data values)
Slightly more likely to detect a statistical difference or relationship if present (if data are normally distributed)	Slightly less likely to detect a statistical difference or relationship if present (if data are normally distributed)

There are other, more specific, assumptions for particular tests, irrespective of whether they are parametric or nonparametric. The assumptions required for the validity of parametric tests tend to be more stringent. Generally, if the assumptions are met then parametric tests should be used, since these have greater **power** than do nonparametric tests. If it is not possible to meet the assumptions, then nonparametric tests are usually appropriate; there are equivalent nonparametric tests available for most parametric tests. Most of the assumptions listed in Table 3.1 are easy to check (e.g. the type of data – if the data are ordinal then we would use a nonparametric test). Testing for normality (whether data are normally distributed) is more complex. There are methods which compare the observed data with a theoretical distribution based on the sample mean, standard deviation and number of sample points – see texts such as Zar (1999) for further details. An easier rule of thumb is based on the spread of data around the mean. The data can be judged to be approximately normally distributed if they satisfy all of the conditions below:

- the frequency distribution of the data looks similar to a normal distribution (i.e. is approximately symmetrical and has only one peak);
- about 68% of the data points lie in the range $\bar{x} \pm s$ (the mean plus or minus the standard deviation);
- almost all of the data lie within the range $\bar{x} \pm 3s$.

Worked Example 3.1 shows this assessment employed on the sulphur dioxide in rainfall data, which are found to be approximately normally distributed. In such cases, as long as the other assumptions are met, parametric statistical tests may (and perhaps should) be used. Otherwise the data should either be transformed (to make the distribution normal – see later in this chapter) before using parametric statistical tests or nonparametric tests should be employed.

WORKED EXAMPLE 3.1 *Assessing conformity to a normal distribution for sulphur dioxide in rainfall*

Using the quick method of assessing normality of data, first produce a frequency histogram from the frequency data (see Table 2.1) and check that the distribution is approximately symmetrical and unimodal.

Frequency table

x (mg)	f
0.5	1
0.6	2
0.7	4
0.8	6
0.9	3
1.0	3
1.1	0
1.2	1

Frequency histogram

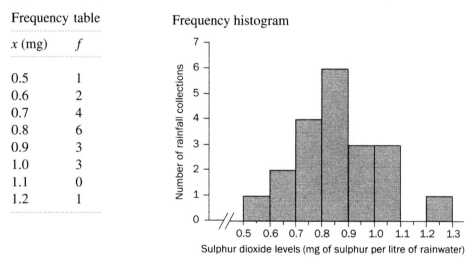

We can see from the frequency histogram that the distribution is approximately symmetrical and unimodal.

Next calculate the mean and standard deviation (see Boxes 2.1 and 2.3). Then work out the range $\bar{x} \pm s$ and see whether approximately 68% of the data points lie within it. From Worked Example 2.1:

the mean = 0.81 mg

the standard deviation = 0.165 mg

So

$$\bar{x} \pm s \quad = (0.81 - 0.165) \text{ to } (0.81 + 0.165)$$

$$= 0.645 \text{ mg to } 0.975 \text{ mg}$$

continued . . .

Worked Example 3.1 continued . . .

We have to be a little careful about how to assess where the data points lie in relation to this range. Remember that data when measured to 0.1 in reality lie in the range of 0.05 to 0.15 (see Chapter 1). So those measurements of 0.6 in the frequency table above could be anything from 0.55 to 0.65, and those measurements of 1.0 could be from 0.95 to 1.05. This means that at least 13 out of 20 (65%), and up to 18 out of 20 (80%), of the data points lie between $\bar{x} \pm s$.

Then work out the range $\bar{x} \pm 3s$ and see whether almost all of the data points lie within it. From Worked Example 2.1:

$$\bar{x} \pm 3s = (0.81) - (3 \times 0.165) \text{ to } (0.81) + (3 \times 0.165)$$
$$= (0.81 - 0.495) \text{ to } (0.81 + 0.495)$$
$$= 0.315 \text{ mg to } 1.305 \text{ mg}$$

We can see from the frequency table above that all of the data points lie between $\bar{x} \pm 3s$.

So, the distribution is reasonably symmetrical, has a single peak, around 68% of the data points lie within the range $\bar{x} \pm s$, and all the data lie within the range $\bar{x} \pm 3s$. Since these data satisfy all of our conditions for normality, we can conclude that the sulphur dioxide in rainfall data are approximately normally distributed.

Transforming data

Variables that are measured on continuous scales (i.e. measurement data: see Chapter 1) are often normally distributed. However, if data are not normally distributed, the data set should be transformed if possible to allow analysis using parametric tests. A transformation is a mathematical operation carried out on each data point within the sample. It is most efficient to transform data using a statistics program or spreadsheet on a computer, because the formula can be entered once and then applied to the whole column. Certain types of data (e.g. derived variables such as ratios, percentages and other proportions, and count data), are particularly likely to fail the normality criteria because their distributions are often skewed (see Chapter 2 and Figures 3.3a, 3.4a and 3.5a). There is an additional problem with count data even if they look symmetrical: that is, the variance can be heavily dependent on the mean (i.e. as the mean varies so too does the variance). This poses a problem for some parametric tests, where an assumption (in addition to the assumption that the data are normally distributed) is that the variances of the different samples are approximately equal (which is less likely to be the case if the variance varies with the mean). Transformations can remove this dependency of the variance on the mean, thus stabilising the variance.

It may seem that using transformations is akin to performing mathematical trickery to manipulate the data. This is not so. In fact, our use of linear measurement scales is in itself arbitrary and indeed, some commonly used environmental variables are already on non-linear scales (e.g. pH and decibels are on logarithmic scales, and area is in squared units). Some of the more commonly used transformations, and the circumstances under which they are used, are described next. Having transformed data, it is important to check (using the guidelines given earlier: see Worked Example 3.1) that they then conform to a normal distribution. If none of the transformations produces an approximately normal distribution, then the data should be analysed using a nonparametric test.

Square root transformation

The square root transformation is used with data that are slightly positively skewed (i.e. with the tail extending to the right: Figure 3.3a), when the variance of a sample is approximately equal to the mean, and/or when the data may conform to a **Poisson distribution** – based on count data where the items counted are relatively rare and are distributed randomly in space or time: see Zar (1999) or Sokal and Rohlf (1995) for further details of the Poisson distribution. To carry out a square root transformation, simply take the square root of each value. If there are zero counts in the data set, then the transformation is improved by adding 0.5 to each data point before taking the square root. Figure 3.3b shows the results of such a transformation on the data displayed in Figure 3.3a.

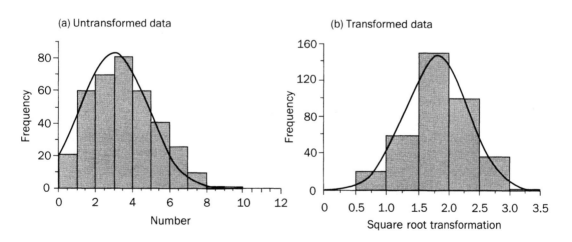

Figure 3.3 *Effects of using a square root transformation on positively skewed count data where the variance is approximately equal to the mean. The fitted curves are the normal distribution curves calculated from the mean and standard deviations*

Logarithmic transformation

This transformation is used when there is a strong positive skew and/or for count data where the variance is greater than the mean (Figure 3.4a: note that the skew is stronger here than in the previous example). The transformation not only produces a more normalised distribution, but also removes the dependence of the variance on the mean (Figure 3.4b). To carry out the transformation, simply take the logarithm of each data point. Since there is no real value to log 0, if there are zero counts in the data, add 1 to all the data points before taking the logarithm.

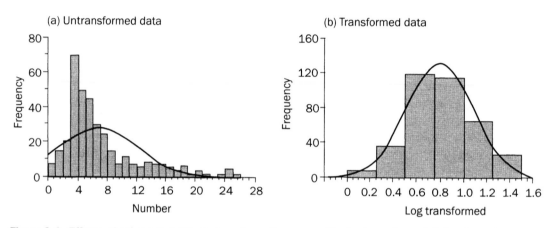

Figure 3.4 *Effects of using a logarithmic transformation on positively skewed count data where the variance is greater than the mean. The fitted curves are the normal distribution curves calculated from the mean and standard deviations*

Other transformations to correct skewness

If you do not know which transformation to use, then there is a series of transformations, which vary in strength, for distributions with varying degrees of skewness. First try the mildest transformation, and if it does not normalise the data (you can assess normality using the method in Worked Example 3.1), move on to the next one. In order of increasing severity of positive skewness (i.e. distributions where the tail extends to the right), try in turn the transformations: $\sqrt{x}$; log x; $1/x$; $1/x^2$. For increasingly negatively skewed distributions (i.e. distributions where the tail extends to the left), try in turn: x^2; x^3; antilog x. Before the use of computers, this would have been a very laborious process. Now it is comparatively easy to try several transformations and generate frequency histograms which may then be examined for normality. If you are calculating by hand, a quicker method of identifying an appropriate transformation is to find the median, quartiles and minimum and maximum points and plot these on a box and whisker plot (see Chapter 2). Then transform only these five numbers and redraw the box and whisker plot. An appropriate transformation is one which produces a symmetrical box and whisker plot. If you can identify such a transformation using these five numbers, then the whole sample can be transformed.

Arcsine transformation

Arcsine transformation is used for proportions (or percentages) where, especially if the data are grouped near to one end or the other of the distribution, skewness or truncation of the data occurs (i.e. an abrupt cut off at 0 or 100%: see Figure 3.5a). If all your data lie between 30% and 70%, then the transformation is probably not necessary, since the data set will not be too truncated. To transform percentages, each value is first divided by 100 (to obtain proportions) then the square root is taken. Finally the inverse sine of each number is found (often seen as $\sin^{-1}$ on a pocket calculator, or arcsine on a computer program). In other words, we are finding the angle whose sine equals the square root for each of the data points. An example of a transformed data set is given in Figure 3.5b. Since this transformation appears a little complex, it is worth working through an example using a single value of 90%. To arcsine-transform 90%, we divide it by 100 to get a proportion (= 0.9), take the square root (= 0.9487) and finally take the inverse sine (= 71.57°). Note that many computer programs calculate the arcsine in radians rather than degrees (degrees are the usual calculating mode for a pocket calculator). You can tell if your computer has worked in radians (as did the program which produced Figure 3.5b) because the values will lie in the range between 0 and 1.57 (where an original value of 100% becomes 1.57); if it has worked in degrees, the values will range between 0 and 90 (where 100% becomes 90). It does not matter whether the transformation is done in degrees or radians, as long as you remain in the same calculator mode to do the back-transformations (see the next section).

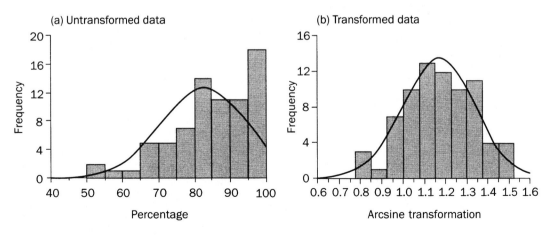

Figure 3.5 *Effects of using an arcsine transformation on percentage data. The fitted curves are the normal distribution curves calculated from the means and standard deviations*

Back-transformation of descriptive statistics

Once data have been transformed successfully, the mean of the transformed data should be used as the measure of central tendency. However, you may wish to present the means and a measure of variation in the scale of the original data. This requires the back-transformation of the statistics. To back-transform the mean value, first calculate the mean of the transformed data, and then back-transform the result:

- for square root transformations use the square of the number (then subtract 0.5 if this was added to each data point during the transformation);
- for logarithmic transformations use the antilog (then subtract 1 if this was added to each data point during the transformation);
- for arcsine transformations obtain the sine and then square this number (then multiply by 100 if percentages were originally used). If the transformed values are in radians, and you are using a calculator for the back-transformation calculations, switch to the 'RAD' mode on the calculator before doing the back-transformation.

Obtaining back-transformed measurements of variation of the original data is slightly more complex. This is because, in any transformation, the variation around the mean of the transformed (normally distributed) data is symmetrical, but when back-transformed it is asymmetrical. The correct procedure is to calculate the 95% upper and lower confidence limits (known as L_{upper} and L_{lower} respectively) of the transformed data (Box 3.3). Worked Example 3.2 shows how to calculate the back-transformed 95% confidence limits for the transformed data shown in Figure 3.3b.

Choosing a statistical test

Having considered whether data are normally distributed, we now need to decide which of the basic statistical tests we require. The tests covered in this book fall into three broad categories:

1 Tests for differences between two (or more) means or medians. These tests are used when you have measurements of one variable that is either continuous (e.g. activity of strontium-90), counts (e.g. numbers of people) or ordinal (e.g. amount of weathering of building stones on a ranked scale), and the data can be divided into two (or more) discrete categories (e.g. clean and polluted sites). There are parametric tests for the analysis of normally distributed variables (or suitably transformed data, e.g. from count or percentage data), and nonparametric equivalents for ordinal or non-normal data. To compare the means or medians of two samples, see Chapter 4; to compare more than two samples see Chapter 7.

2 Tests for a relationship between two (or more) variables. These tests are appropriate when you have obtained two (or more) measurements of continuous,

---

Box 3.3 *Calculating back-transformed 95% confidence limits of the mean*

To provide measures of variation around the mean for back-transformed data, the 95% confidence limits are calculated from the transformed data using the following formula (for $n < 30$):

$$95\% CL_{\text{trans}} = t_{0.05[n-1]} SE_{\text{trans}}$$

where:

$95\% CL_{\text{trans}}$ are the 95% confidence limits of the transformed data;

$t_{0.05[n-1]}$ is the value from a table of the t statistic at $P < 0.05$ and df $= n - 1$ (see Table D.2, Appendix D);

SE_{trans} is the standard error of the transformed data.

For $n \geq 30$, we can use the z value of 1.96 instead of the t value (see Chapter 2), so:

$$95\% CL_{\text{trans}} = 1.96\, SE_{\text{trans}}$$

The upper and lower 95% confidence limits ($L_{\text{upper trans}}$ and $L_{\text{lower trans}}$) are then calculated as:

$$L_{\text{lower trans}} = \bar{x}_{\text{trans}} - 95\% CL_{\text{trans}}$$

and:

$$L_{\text{upper trans}} = \bar{x}_{\text{trans}} + 95\% CL_{\text{trans}}$$

where $\bar{x}_{\text{trans}}$ is the mean of the transformed data.

Finally, the 95% confidence limits are back-transformed in the same way as for the means. These asymmetric limits around the mean may then be quoted in the results text or tables, or plotted as error bars on appropriate bar or point charts.

---

count or ordinal data for each individual sampling unit. A sampling unit is a statistical individual, for example an individual household from which the number of people and amount of water consumed have been recorded, or an individual site for which you have the concentration of a pollutant and a measure of the number of species present. There are different tests depending on: whether there is a linear (rather than a curved) relationship between the two variables; whether you are simply determining whether there is a relationship or trying to predict the values of one variable from another; and whether your data are interval/ratio (including count data) or ordinal. See Chapter 5 for how to choose a suitable test.

WORKED EXAMPLE 3.2 *Back-transformed means and 95% confidence limits for square root transformed data*

From the square root transformed data used to create Figure 3.3b, the following statistics have been obtained:

$$\bar{x}_{trans} = 1.816$$
$$SE_{trans} = 0.026$$
$$n = 371$$

Using the formula in Box 3.3 for $n \geq 30$, we can calculate the confidence limits as

$$95\%CL_{trans} = 1.96SE_{trans} = 1.96 \times 0.026 = 0.050\,96$$

We then calculate the upper and lower transformed limits as

$$L_{lower\;trans} = \bar{x}_{trans} - 95\%CL_{trans} = 1.816 - 0.050\,96 = 1.765\,04$$

$$L_{upper\;trans} = \bar{x}_{trans} + 95\%CL_{trans} = 1.816 + 0.050\,96 = 1.866\,96$$

These are then back-transformed, remembering to subtract 0.5 from each point since 0.5 was originally added to correct for having zero values (where zeros do not occur in the original data, this subtraction is not needed):

$$\bar{x}_{back\text{-}transformed} = \bar{x}_{trans}^{2} - 0.5 = 1.816^{2} - 0.5 = 3.297\,857 - 0.5 = 2.797\,856$$

(note that this is different from the mean of the original data, which would have been 3.049).

The confidence limits may also be back-transformed in a similar way

$$L_{lower\;back\text{-}transformed} = L_{lower\;trans}^{2} - 0.5 = 1.765\,04^{2} - 0.5 = 3.115\,37 - 0.5 = 2.615\,37$$

$$L_{upper\;back\text{-}transformed} = L_{upper\;trans}^{2} - 0.5 = 1.866\,96^{2} - 0.5 = 3.485\,54 - 0.5 = 2.985\,54$$

It is apparent that the back-transformed mean does not lie exactly between these limits, but is slightly closer to the lower one. This asymmetry in the distribution is confirmed if we inspect the original data (Figure 3.3a).

3 Frequency analysis. So far in our consideration of the analysis of frequency (count) data, we have concentrated on transforming the data so that they can be treated as if they were normally distributed. For example, in a quadrat survey of a field, if the number of plants found within several quadrats were counted, a distribution such as that in Figure 3.3 might be generated (which we could transform prior to appropriate analysis). However, imagine we wanted to compare the number of hedgehogs found in rural gardens with the number found in suburban gardens. In theory, we could use the same survey technique, taking several readings over a large area. In practice, this would be very difficult: the quadrats would have to be large, and the total area sampled massive. The other approach is to look at just one fairly large equally sized area for each type of garden. This would generate just two data points: the number of hedgehogs in rural gardens, and the number in suburban gardens. Using only these two data points, we could test the null hypothesis that there were equal numbers in each garden type. This sort of frequency analysis is known as a goodness of fit test, and is used to test observed distributions against expected distributions. The other application of frequency analysis is the test of association. In the hedgehog example, the counts were divided up depending on one category: type of garden. If we had also classified each hedgehog by another category, e.g. as adult or juvenile, we could discover whether the relative numbers of juveniles and adults were influenced by garden type (i.e. whether there was an association between garden type and hedgehog age). Note that this example is similar to the count data component of cross-tabulation (see Chapter 2). See Chapter 6 for suitable tests.

A key to the tests covered is given at the very end of the book.

Summary

By now you should be able to:

- formulate the null hypothesis;

- decide whether there is a significant result from a statistical test, based on the P value;

- assess the normality of data, visually (for symmetry and a single peak) and by checking that about 68% of the data points lie in the range $\bar{x} \pm s$ and that almost all of the data points lie in the range $\bar{x} \pm 3s$;

- transform data using logarithmic, square root and arcsine transformations;

- back-transform means and confidence limits calculated on the transformed data;

- decide which sort of analysis question you want to ask (e.g. comparing means or medians, exploring relationships, or frequency analysis).

Questions

3.1 In a survey to assess the extent of fungal infection of a certain tree species, the percentage of leaves showing any infection was calculated for each of 20 individual trees. The following data (percentages) were recorded:

10 2 4 5 19 2 1 5 5 4 5 8 12 4 8 4 4 5 9 8

(i) Plot the data as a frequency histogram.
(ii) Arcsine-transform the data and plot the transformed data.
(iii) Calculate the following statistics on the transformed data (calculations may use either degrees or radians). See Table D.2 (Appendix D) for t values (for confidence level calculations).

Mean	
Standard deviation	
Standard error	
Degrees of freedom	
Lower 95% confidence level (L_{lower})	
Upper 95% confidence level (L_{upper})	

(iv) Back-transform the following statistics:

Mean	
Lower 95% confidence level (L_{lower})	
Upper 95% confidence level (L_{upper})	

3.2 The size and layout of urban areas influence their climatic conditions. A survey was designed to determine whether there is a difference in the mean maximum temperatures recorded within an urban area and in an equivalent rural site.

(i) Formulate the null hypothesis.
(ii) The maximum temperature was recorded each day for one month. The mean for the urban area was 26.0°C, whereas the mean for the adjacent rural area was 24.5°C. A statistical test gave the probability of the null hypothesis being true as 0.045. Should you accept or reject the null hypothesis?
(iii) What would you conclude about the effect of urbanisation on the maximum temperatures recorded?

4　Differences between two samples

When we examine a potential difference between two samples, the test chosen depends on the nature of the data (i.e. whether they are normally distributed or not, and whether they are in pairs or not). This chapter covers:

- the *t* test for normally distributed, unmatched data
- the Mann–Whitney *U* test for unmatched data measured at least on an ordinal scale
- the paired *t* test for normally distributed differences between matched data
- the Wilcoxon matched-pairs test for differences between matched data measured at least on an ordinal scale

This chapter covers how to determine whether there is a statistically significant difference between the central tendency of two samples (there are equivalent tests both for normal and for ordinal or non-normal data). The broad type of test we use depends upon whether the data are **matched** or **unmatched**. Let us look again at our example from Chapter 1 where we were interested in the lead levels in cabbages grown in allotments near to and far away from major roads. We could measure the lead levels in cabbages from allotments which we can identify as being near to major roads and those from allotments far away from major roads and produce the data shown in Table 4.1a. We can see that these data are unmatched. That is, the lead level from cabbages from the first allotment near to a major road (0.04 mg per kilogram) is not linked in any way to that from the first allotment far away from the major road (with a value of 0.03 mg per kilogram). We would analyse such data using tests for unmatched comparisons.

An alternative survey design would have been to identify pairs of allotments, one of which is near to a particular road, the other being further from the same road (and not near to any other major road). The data in Table 4.1b are paired for a given road. For example, the allotment near to Cardiff Road is linked to one further away from Cardiff Road, therefore these are matched samples. We would analyse such data using tests for matched comparisons which allow us to examine differences between the allotments near to and far from roads while taking into account differences between the roads. For example, Cardiff and London Roads seem to have higher levels of lead. Such differences could be due to the traffic rates along the roads, or differences in the local environment (such as the presence of naturally occurring lead) or local climatic conditions (which may influence the dispersal of lead).

Table 4.1 *Comparison of unmatched and matched data for lead levels (mg kg⁻¹) in cabbages from allotments near to and far from major roads*

(a) Unmatched data		(b) Matched data		
Near to a major road	Far from a major road	Road	Near to the road	Far from the road
0.04	0.03	Cardiff Road	0.07	0.08
0.05	0.03	Edinburgh Road	0.06	0.04
0.06	0.06	London Road	0.09	0.07
0.06	0.05	Manchester Road	0.05	0.04
0.04	0.05	Oxford Road	0.06	0.05
etc.	etc.	etc.	etc.	etc.

Unmatched comparisons

Consider our example from Chapter 3 where we examined strontium-90 activity in milk from farms near to two types of nuclear installation. This is an unmatched design, because each farm is independent of every other farm. Imagine there was very little variation in the activity of strontium-90 in the milk from farms near to Power Gush and Rayzoom plants and they had different levels of activity. If we estimated parameters (i.e. the mean and standard deviation) using the sample data for each farm location type, then their ideal normal distributions might look like those in Figure 4.1. Clearly there is no overlap at all between the two samples and so the populations are different. However, in reality, environmental data are usually much more variable and our ideal normal plots are more likely to look like those in Figure 4.2a.

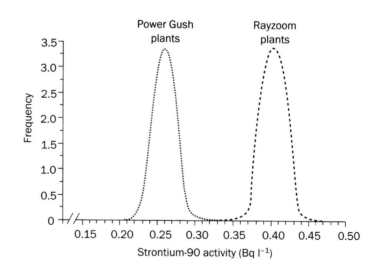

Figure 4.1 *Normal distributions of strontium-90 activity in milk from farms near to two types of nuclear plant: non-overlapping distributions*

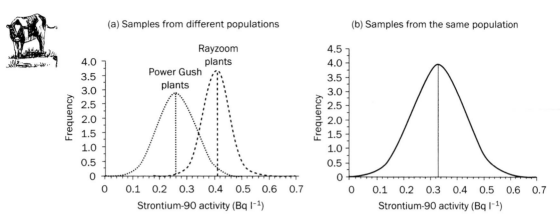

Figure 4.2 *Normal distributions of strontium-90 activity in milk from farms near to two types of nuclear plant: overlapping distributions*

The milk from farms near to the Power Gush plants has the lower mean value of 0.263 Bq l⁻¹, whereas that from farms near to the Rayzoom plants is on average 0.411 Bq l⁻¹. However, we can see considerable overlap between these samples. The question we wish to ask is whether these samples really come from two different populations with these different mean values, or whether they in fact come from the single population illustrated in Figure 4.2b. To determine which scenario is more likely (i.e. whether these two samples differ significantly) we need to use a statistical test with the null hypothesis that the samples do come from the same population (i.e. that illustrated in Figure 4.2b).

In a situation such as above, where data from one sample (e.g. farms near Power Gush plants) are independent of the data in the other sample (e.g. farms near Rayzoom plants), we use a statistical test for unmatched (sometimes called unpaired or independent) samples. For normally distributed data we use a *t* test, and for non-normal or ordinal data we use a **Mann–Whitney *U* test**. Of course, we could have a situation where we wished to compare more than two samples (perhaps if we also had a sample of farms near to a third type of nuclear power plant). A suite of statistical tests to analyse such data is presented in Chapter 7.

t test

If the distributions for the two samples are normal (as is the case for the strontium-90 in milk example) and if the sample sizes are small (under 30 individual sample units in either sample) then a *t* test is appropriate. This test (which is sometimes called Student's *t* test or an independent sample *t* test) compares the **absolute difference** between two means (i.e. we are not concerned at this point with which mean is larger, only with the magnitude of difference involved).

The difference between means provides us with a measure of the explained variation between the samples, i.e. the variation caused by the category (here, the different nuclear plant types). This difference is adjusted to take into account the amount of unexplained variation within each sample (measured using the standard deviations) and the sample size (the number of each type of farm). Large standard deviations make it less likely that a statistically significant difference between means will be found because they imply that there is a lot of variability that is not accounted for by membership of the categories. Large sample sizes allow us to be more confident in our approximations of the population means and standard deviations, and so reduce the impact of the standard deviations.

There are several versions of the formula for calculating the t statistic (sometimes called t_s), most of which are user-friendly approximations of a quite complex-looking fuller formula (see Box 4.1). Although it looks a little daunting, all of the numbers required are reasonably familiar (means, standard deviations and sample sizes). Many statistical computer programs use the full formula.

Figure 4.3 shows the relationship between the explained variation, unexplained variation, t value, probability and the null hypothesis. Larger t values lead to smaller P values, i.e. they are more likely to be significant and lead to a rejection of the null hypothesis. If, on the other hand, t is small then the null hypothesis that there is no significant difference between means is more likely to be true. Just how large the t value needs to be for significance depends on the degrees of freedom, which in this case are $n_1 + n_2 - 2$, where n_1 is the sample size of sample 1 and n_2 is the sample size of sample 2. Once calculated, we find the probability of obtaining the test statistic (t)

Box 4.1 Formula for the t statistic

The formula for t is:

$$t = \frac{|\bar{x}_1 - \bar{x}_2|}{\sqrt{\dfrac{(n_1 - 1)\ s_1^2 + (n_2 - 1)\ s_2^2}{(n_1 + n_2 - 2)} \times \dfrac{n_1 + n_2}{n_1 n_2}}}$$

where:

$\bar{x}_1$ and $\bar{x}_2$ are the means for samples 1 and 2, respectively;

s_1^2 and s_2^2 are the variances of samples 1 and 2 (i.e. the square of the standard deviation; see step 3 of Box 2.3);

n_1 and n_2 are the sample sizes of samples 1 and 2.

The symbol |...| instructs you to take the absolute value of the term within the lines (i.e. ignoring the sign so that negative values become positive). Note that as the difference between means gets larger (the top line of the formula), and as the standard deviations (on the bottom line of the formula) get smaller, the t value increases. To recap: the bigger the difference between the means (explained variation), and/or the smaller the standard deviation (unexplained variation), the larger the t value.

by chance. To do this we compare our calculated value of t to a table containing t values with their associated probabilities of the null hypothesis being true. This involves looking up the value you obtain from the equation in Box 4.1 in a table of the t distribution (an extract of which is shown in Table 4.2) against the appropriate degrees of freedom. This gives a probability (P) of obtaining the calculated t value by chance, i.e. the probability of there being no significant difference between the means. Only if this P value is lower than 0.05 can we reject the null hypothesis and state that there is a significant difference between the means. Most computer programs calculate the probability for you, negating the need to consult tables.

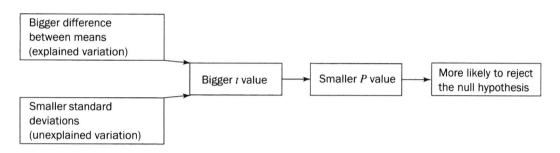

Figure 4.3 *Relationship between the components of the t test and the null hypothesis*

Worked Example 4.1 goes through the steps of the calculation of t for the strontium-90 example. Although it is preferable when designing a project to try to keep the sample sizes equal, it is not always possible. Here, for example, we have only nine Rayzoom plants, compared to ten Power Gush plants. This type of situation frequently occurs in environmental data gathering where one category is represented less frequently than the other. Fortunately, equal sample sizes are not essential for the t test. For these data, t is 5.413. The degrees of freedom are 17 ($n_1 + n_2 - 2 = 10 + 9 - 2$). Knowing the degrees of freedom, we look up our t value in a table of t values (see the shaded section of Table 4.2), where we are looking for a calculated t value which is greater than the table value of t if we are to reject our null hypothesis. Here, we find that the t value we have (5.413) is higher than the table t value at df = 17 for both $P = 0.05$ (table $t = 2.110$) and $P = 0.01$ (table $t = 2.898$). Note that for any specific number of degrees of freedom, as the probability decreases, the t value in the table increases (i.e. for any given degrees of freedom, the higher the t value, the less likely it is to have arisen by chance). Most, but not all, test statistics are distributed so that probabilities decrease with increasing values of the statistic; you need to examine each table with care. Since our calculated t value is greater than 2.898 (the value at $P = 0.01$) the probability of obtaining the t value we have by chance is less than 0.01 which implies a low probability of there being no significant difference between the means. Since P is lower than the critical level of 0.05, we reject the null hypothesis of no significant difference and accept an alternative hypothesis that there is a significant difference.

WORKED EXAMPLE 4.1 *Calculating a t test statistic for strontium-90 activity in milk from farms near to two types of nuclear plant*

Power Gush plants		Rayzoom plants	
($n_{PG} = 10$)		($n_R = 9$)	
x_{PG}	x^2_{PG}	x_R	x^2_R
0.25	0.0625	0.43	0.1849
0.34	0.1156	0.35	0.1225
0.26	0.0676	0.42	0.1764
0.20	0.0400	0.39	0.1521
0.21	0.0441	0.45	0.2025
0.15	0.0225	0.49	0.2401
0.27	0.0729	0.34	0.1156
0.26	0.0627	0.39	0.1521
0.31	0.0961	0.44	0.1936
0.38	0.1444		

$\sum x_{PG} = 2.63$ $\sum x^2_{PG} = 0.7333$ $\sum x_R = 3.70$ $\sum x^2_R = 1.5398$

$\bar{x}_{PG} = 0.263$ $\bar{x}_R = 0.411\,111\,11$

$s^2_{PG} = 0.004\,623\,333$ $s^2_R = 0.002\,336\,111\,1$

First calculate the mean ($\bar{x}$) and variance (s^2) of the two samples (these are given below the columns above: see Boxes 2.1 and 2.3 for how to calculate these). Use the mean, variance and the sample sizes in the formula below to calculate t:

$$t = \frac{\left| \bar{x}_{PG} - \bar{x}_R \right|}{\sqrt{\frac{(n_{PG} - 1)\, s^2_{PG} + (n_R - 1)\, s^2_R}{(n_{PG} + n_R - 2)} \times \frac{n_{PG} + n_R}{n_{PG} n_R}}}$$

$$= \frac{\left| 0.236 - 0.411\,111\,11 \right|}{\sqrt{\frac{(10 - 1) \times 0.004\,623\,3 + (9 - 1) \times 0.002\,336\,1}{(10 + 9 - 2)} \times \frac{10 + 9}{10 \times 9}}}$$

$$= \frac{0.148\,111}{\sqrt{0.003\,546\,811\,765 \times 0.211\,111\,111}}$$

$$= \frac{0.148\,111}{0.027\,364} = 5.413$$

continued . . .

Worked Example 4.1 continued . . .

Calculate the degrees of freedom:

$$df = n_{PG} + n_R - 2 = 10 + 9 - 2 = 17$$

Look up t in the table at df = 17. The table value is 2.110 for $P = 0.05$ and 2.898 for $P = 0.01$. Our value of 5.413 is greater than either of these, therefore we can reject the null hypothesis of no difference and accept an alternative hypothesis that there is a difference between the means ($P < 0.01$). Looking at the means, strontium-90 activity in milk is significantly higher from farms near to Rayzoom plants than it is from those near to Power Gush plants.

Table 4.2 Selected values of t. Calculated values of t greater than the table values are significant. Shaded areas indicate the critical values for the example referred to in the text. A more comprehensive table of t values is given in Table D.2 (Appendix D)

df	t	
$(n_1 + n_2 - 2)$	**P = 0.05**	P = 0.01
14	**2.145**	2.977
15	**2.131**	2.947
16	**2.120**	2.921
17	2.110	2.898
18	**2.101**	2.878

You need to record the probability, t value and number of degrees of freedom, together with a statement to the effect that there is a significant difference between the samples. Although we know that there is a significant difference, we do not yet know where that difference lies; you need to examine the sample means to see which is the larger (in this example the mean for farms near to Rayzoom plants is greater than that for farms near Power Gush plants). In the results section of a report, you might say:

Milk from farms near to Rayzoom plants has significantly higher strontium-90 activity than does milk from farms near to Power Gush plants ($t = 5.413$, df = 17, $P < 0.01$).

You should also give an indication of the means ($\pm$ standard errors), either in the text, in a table, or in a bar chart or point chart (see Chapter 2).

Under some circumstances, some computer programs calculate the degrees of freedom for t tests as slightly different from $n_1 + n_2 - 2$. This is because one assumption of the t test is that the variation within one sample is of a similar magnitude to that in the other sample. In cases where the variances are not equal, some programs alter the degrees of freedom to correct for this. If your program has done this, use the corrected degrees of freedom and associated probability. In all cases, it is the probability (P value) which is the important figure for the interpretation of the results, but always record the degrees of freedom and the test statistic as well. Some computer programs give negative values of t: you should record the absolute value of t (i.e. without the minus sign if present). The reason for the appearance of negative values of t is that the program will have used the difference between the means, rather than the absolute difference between the means, when calculating t (see the top line of the formula in Box 4.1). Whether the sign is positive or negative is arbitrary, and depends purely on whether the first mean entered into the formula was greater or smaller than the second mean.

The data in Worked Example 4.1 are displayed in two separate columns (one for each sample). This is for clarity of display. Note that, if you use a computer program to calculate your t tests, this data format will not usually be appropriate. See Table B.1 (Appendix B) for details of data entry for statistical analysis using computers.

It is worth noting that there is also a one-sample t test which allows comparisons of a single mean value with an expected value – see Zar (1999) or Sokal and Rohlf (1995) for further details. Be careful when using computer programs that the test you are using is called a two-sample test or a test for comparisons between means.

Where the samples are large (usually over 30 in each sample) a z **test** can be used instead of a t test. The equation is similar to that used in the t test, although the calculation is easier (see Box 4.2).

Box 4.2   *Formula for the z statistic*

When both sample sizes are high (> 30), a z test is used as an alternative to a t test. z is calculated as:

$$z = \frac{\left| \bar{x}_1 - \bar{x}_2 \right|}{\sqrt{\dfrac{s_1^2}{n_1} + \dfrac{s_2^2}{n_2}}}$$

where:

$\bar{x}_1$ and $\bar{x}_2$ are the means for samples 1 and 2, respectively;

s_1^2 and s_2^2 are the variances of samples 1 and 2 (i.e. the square of the standard deviation, see step 3 of Box 2.3);

n_1 and n_2 are the sample sizes of samples 1 and 2.

The test statistic is compared to values in a z table (see Table D.1 in Appendix D: also called a table of standardised normal deviates). Since z values have no degrees of freedom, the table is very simple and values of 1.96 or less indicate a probability greater than or equal to 0.05 (i.e. no significant difference), whilst values greater than 2.576 indicate a probability less than 0.01 (i.e. a highly significant difference between the samples). z values may also be looked up in t tables at degrees of freedom of infinity (see Table D.2).

Mann–Whitney *U* test

The Mann–Whitney U test (also sometimes called the Wilcoxon–Mann–Whitney test) provides a nonparametric equivalent to the t test that allows comparisons of samples which are not normally distributed (although even this test assumes that the data are from broadly similar distributions; i.e. it should not be used if the samples are strongly skewed in different directions). Mann–Whitney U tests can be applied to normally distributed data, but in this case will be less powerful than a t test (i.e. there is a higher likelihood of accepting the null hypothesis when in fact it is false – committing a type II error). In the Mann–Whitney U test, the measure of central tendency being compared is the median rather than the mean. Like many nonparametric tests, all calculations are performed on the rank position of each data point, rather than the actual numbers. Although this conversion of data to ranks loses some of the precision in the data, it does allow comparisons between samples which have been measured on an ordinal scale as well as those measured on interval or ratio scales.

When calculating the Mann–Whitney U test by hand, the first stage is to convert the data to rank values. The data points in both samples are compiled together in a single list and placed in order (preferably ascending order to prevent confusion). Each data point is then given a **ranked value** based on its position in the overall order, where the first value is given the rank of 1, the next the rank of 2, and so on. The rank values are then summed separately for each sample (to give $\sum R_1$ and $\sum R_2$, the sum of the ranks of samples 1 and 2 respectively) and the test statistics U_1 and U_2 are calculated using the formulae in Box 4.3. Whichever value of U is lower is then compared to the values in a table of U values (an extract of which is

---

Box 4.3   *Formula for the Mann–Whitney U statistic*

Two values of U are calculated as follows:

$$U_1 = n_1 n_2 + \frac{n_1(n_1 + 1)}{2} - \sum R_1$$

and

$$U_2 = n_1 n_2 + \frac{n_2(n_2 + 1)}{2} - \sum R_2$$

where:

  n_1 and n_2 are the numbers of data points in samples 1 and 2 respectively;

  $\sum R_1$ and $\sum R_2$ are the sums of the ranks in samples 1 and 2.

If the arithmetic has been correctly performed then

  $U_1 + U_2 = n_1 n_2$ (check this is so)

The smaller of the two U values is chosen for comparison with the appropriate statistical table (Table D.3 in Appendix D).

shown in Table 4.3). Although the example that will be used has equal sample sizes, this test can also be applied to samples of different sizes.

Imagine a survey where tree condition is compared between polluted and clean sites. The leaves of many plant species show colour changes with stress. Leaf condition could be measured on the following scale, where a low score indicates leaves in poor condition:

6 = leaves dark green all over
5 = leaves mainly dark green but some have lighter speckling
4 = several leaves have lighter speckling
3 = many leaves have patches of light colour
2 = most leaves have large patches of light colour
1 = most leaves have major areas of light colour

If we examine the largest tree in each of ten clean and ten polluted sites, our data could be as follows:

Clean   4 5 4 4 5 6 6 6 6 3

Polluted   2 2 2 1 6 4 4 5 4 3

We first sort the data from both samples in a single column, place them in ascending order, and give them rank values (see Worked Example 4.2). Note that **ranking data** becomes slightly more complicated if two or more data points have the same value. When this occurs we need to calculate the tied rank for those points with equal values (see Box 4.4 and Worked Example 4.2). We can see that, when placed in rank order, most of the trees with low scores (i.e. in poor condition) are from the polluted sites (shaded in this worked example), whilst most of the trees in good condition come from clean sites (and occur nearer the bottom of the table). This implies that the trees in the clean site have higher scores than those from the polluted site. However, there is some overlap between the condition scores from the clean and polluted sites. To find if the apparent difference is significant, the ranks can now be summed for each sample, and the test statistic calculated. From our example, the lowest value of U (21) can be compared with the appropriate table of critical values (see the shaded area of Table 4.3). We can see that where both n values equal 10, the table value is 23 with a probability of 0.05 and 16 with a probability of 0.01. Our calculated value of 21 lies between these two tabled values, therefore our probability is between 0.05 and 0.01. That is, there is a significant difference ($P < 0.05$) in the condition of trees in the two site types. Note that the smaller the U value, the smaller the probability: this is the reverse of the case when we looked up t values for the t test.

To identify where the difference is, we calculate the median values. The trees on the clean sites have a median of 5, while the plants on the polluted sites have a median

WORKED EXAMPLE 4.2 *Calculating a Mann–Whitney* U *test statistic for tree condition (using ranked colour scores) in clean and polluted sites*

Site type	Condition score	Ascending order	Calculating the rank value	Rank value[a]
Polluted	1	1	1	1
Polluted	2	2	$\dfrac{2+3+4}{3}=3$	3
Polluted	2	3		3
Polluted	2	4		3
Clean	3	5	$\dfrac{5+6}{2}=5.5$	5.5
Polluted	3	6		5.5
Polluted	4	7		9.5
Clean	4	8		9.5
Polluted	4	9	$\dfrac{7+8+9+10+11+12}{6}=9.5$	9.5
Clean	4	10		9.5
Polluted	4	11		9.5
Clean	4	12		9.5
Polluted	5	13		14
Clean	5	14	$\dfrac{13+14+15}{3}=14$	14
Clean	5	15		14
Clean	6	16		18
Clean	6	17		18
Polluted	6	18	$\dfrac{16+17+18+19+20}{5}=18$	18
Clean	6	19		18
Clean	6	20		18

[a] Ranked values from polluted sites are shaded to discriminate between the two site types.

n_{Clean} and n_{Polluted} both equal 10, and the sums of ranks for the two samples are

$$\Sigma R_{\text{Clean}} = 134 \text{ and } \Sigma R_{\text{Polluted}} = 76$$

Next check the ranking accuracy by calculating the sum of all possible ranks in two ways and checking that they agree:

continued . . .

Worked Example 4.2 continued . . .

$$\sum R_{Total} = \sum R_{Clean} + \sum R_{Polluted} = 134 + 76 = 210$$

and

$$\sum R_{Total} = \frac{(n_{Clean} + n_{Polluted}) \times (n_{Clean} + n_{Polluted} + 1)}{2} = \frac{(10+10) \times (10+10+1)}{2} = 210$$

Since these both agree, we are sure that we have ranked our data correctly. We can now calculate both values of U:

$$U_{Clean} = n_{Clean}\, n_{Polluted} + \frac{n_{Clean}(n_{Clean}+1)}{2} - \sum R_{Clean} = (10 \times 10) + \left(\frac{10\,(10+1)}{2}\right) - 134 = 21$$

$$U_{Polluted} = n_{Clean}\, n_{Polluted} + \frac{n_{Polluted}(n_{Polluted}+1)}{2} - \sum R_{Polluted} = (10 \times 10) + \left(\frac{10\,(10+1)}{2}\right) - 76 = 79$$

The smaller of these ($U = 21$) is then compared with the critical value given in appropriate tables (see Table 4.3).

Also check that $U_{Clean} + U_{Polluted} = n_{Clean}\, n_{Polluted}$ ($21 + 79 = 10 \times 10$) to confirm that the U value calculations are accurate.

Table 4.3 *Selected values of* U *for the Mann–Whitney* U *test. For significance, calculated* U *values must be less than or equal to the table value. Shaded lines indicate the critical values for the example referred to in the text. A more comprehensive table of* U *values is given in Table D.3 (Appendix D). The upper table values (in bold) are for P = 0.05, while the lower values are for P = 0.01*

	Larger n value					
Smaller n value	7	8	9	10	11	12
7	**8**	**10**	**12**	**14**	**16**	**18**
	4	6	7	9	10	12
8		**13**	**15**	**17**	**19**	**22**
		7	9	11	13	15
9			**17**	**20**	**23**	**26**
			11	13	16	18
10				**23**	**26**	**29**
				16	18	21
11					**30**	**33**
					21	24
12						**37**
						27

of 3.5. Therefore, we can record the U, n_1, n_2 and P values and say that:

> The trees on the clean sites have a higher condition score than those on the polluted sites ($U = 21$, $n_{Clean} = 10$, $n_{Polluted} = 10$, $P < 0.05$).

Remember also to display the medians and interquartiles, either in the text, a table, or graphically using a box and whisker plot (see Figure 2.11 which has been drawn using this example).

In the case of a significant difference, a faster way to decide which sample has larger scores is to divide the sum of the ranks by the sample size to obtain the mean rank for each sample. The sample with the higher **rank mean** is significantly larger than the other sample.

Note that the data for tree condition in Worked Example 4.2 are listed in a single column, with another column for the site type. This is also likely to be the way to enter data when using a computer program to calculate your Mann–Whitney U tests (see Table B.1 in Appendix B for further details of data entry for statistical analysis using computers).

As the sample size becomes large, the probability distribution of the test statistic U becomes very similar to that for z (which we encountered in Box 4.2). There is a formula to calculate z from U, making it possible to assess the significance by looking up the probability for z instead. Details of this procedure are not covered here – see texts such as Zar (1999) or Siegel and Castellan (1988) – it is mentioned because statistics programs often give z (and its P value) as well as U. If your computer program does this, then you should record U, z and P, as well as n_1 and n_2.

Strictly speaking, when there are tied values (as in Worked Example 4.2, where, for example, several sites had a rank value of 9.5), the P value is slightly inaccurate. This inaccuracy leads to the probability being higher than it should be, thus increasing the likelihood of falsely accepting the null hypothesis (type II error). The more tied ranks there are, the more inaccurate the calculation of the probability. A correction factor can be applied which makes use of the approximation to the z distribution – see tests such as Zar (1999) or Siegel and Castellan (1988). If your computer program gives you corrected values for ties, then record the corrected P and z values. If you are performing calculations by hand, then this inaccuracy is a problem only if you have a marginally non-significant result (i.e. P is just greater than 0.05). If the result is significant, or P is much greater than 0.05, it will remain so even if corrected for ties. Tied ranks may also be taken into account by simply ranking the data and using a t test on the ranks.

Matched-pair tests

When we wish to determine whether there are differences between two means (or medians) of matched data, then we employ paired or matched tests. These allow us to

---

Box 4.4 *Calculating tied ranks*

If we have a data set where several points have the same value, then we must allow for this in the ranking procedure by calculating the tied rank. In the 20 data points below, several have the same value. The data are first ordered to give values from 1 to 20. Data with the same value are given their mean rank order:

Score	1	2	2	2	3	3	4	4	4	4	4	4	5	5	5	6	6	6	6	6
Order	1	2	3	4	5	6	7	8	9	10	11	12	13	14	15	16	17	18	19	20
Ranks	1	3	3	3	5.5	5.5	9.5	9.5	9.5	9.5	9.5	9.5	14	14	14	18	18	18	18	18

You can see that a score of 1 is the lowest value so this is given a rank of 1. There are three scores of 2 so these all receive the rank of 3, calculated as the next three ranks $(2 + 3 + 4)$ divided by the number of items (3). Next we have two values of 3 which are given ranks of 5.5 $(5 + 6$ divided by 2). So in general, tied ranks are given the mean rank of the group of ranks which would have applied had there been no tied values. You can check your accuracy in ranking the data by summing the ranks $(\sum R)$ and checking that:

$$\sum R = \frac{n(n+1)}{2}$$

where n is the number of items being ranked.

Note that when using a computer, the program automatically ranks the data before carrying out the test.

---

find the significance of differences in response whilst taking into account the fact that there may be differences in the initial status of each sample unit. There are parametric (**paired *t* test**) and nonparametric (**Wilcoxon matched-pairs test**) tests for paired comparisons. There is also a suite of statistical tests suitable for situations where we have more than two matched samples (see Chapter 7).

Paired *t* test

The paired *t* test is appropriate when the differences between pairs of data are normally distributed. However, the data themselves do not necessarily have to be normally distributed. The null hypothesis is that there is no significant difference between pairs of measurements and, if this is so, we would expect these differences (*d*) to be clustered around a mean difference of zero. We also take into account the variation in differences, by using the standard error of the mean difference. The test statistic (*t*, sometimes called t_p) is calculated as shown in Box 4.5.

One example where we could use a paired *t* test is if we sampled invertebrates above and below a sewage outflow on each of ten rivers. The number of species of invertebrate is shown in Worked Example 4.3. Had we used an unpaired *t* test, we

---

Box 4.5 *Formula for calculating the* t *statistic for paired* t *tests*

First we find, the difference (d) between each pair of data points. These are summed and the result is then divided by the number of data pairs to find the mean difference ($\bar{d}$):

$$\bar{d} = \frac{\sum d}{n}$$

where:

d is the difference between each pair of data points;

n is the number of pairs of data points.

We then find the standard error of the mean difference (SE_d):

$$SE_d = \sqrt{\frac{\sum d^2 - \frac{\left(\sum d\right)^2}{n}}{n(n-1)}}$$

Finally, we can calculate the statistic (t):

$$t = \frac{\bar{d}}{SE_d}$$

---

would have simply compared the number of species in each sample and found that there was no significant difference between the number found above and below the outflow ($t = 1.31$, df $= 18$, $P > 0.05$). However, as we can see from the data, although the individual rivers vary considerably (e.g. river H only has half as many species as river J), in only one case are there actually more invertebrate species below the outfall. A paired t test on these data (see Worked Example 4.3) accommodates the variation in rivers, giving $t = 2.68$. We look up this value in the same t table as before (see Table D.2, Appendix D), but with the degrees of freedom being calculated as the number of data pairs minus 1 (i.e. the number of rivers minus 1 = 9). Note that the

unpaired t test would assume that there were 20 independent rivers ($20 - 2 = 18$ degrees of freedom). The table value at 9 degrees of freedom and $P = 0.05$ is 2.262. Our value is larger, therefore the probability is less than 0.05, and there is a significant difference in the number of invertebrate species found above and below the outflow. The test statistic (t) can be negative or positive, depending on whether you subtract the smaller readings from the larger (as in Worked Example 4.3) or vice versa. In paired t tests, we can use the sign of t to see which direction the difference is in: if the second sample is subtracted from the first and a positive t value is obtained,

WORKED EXAMPLE 4.3 *Calculating a paired t test statistic for numbers of invertebrate species above and below a sewage outflow*

River	No. of species above outflow	below outflow	d (above minus below)	d²
A	8	6	2	4
B	9	9	0	0
C	12	11	1	1
D	8	4	4	16
E	15	10	5	25
F	7	8	−1	1
G	14	10	4	16
H	5	5	0	0
I	7	6	1	1
J	11	10	1	1
			$\sum d = 17$	$\sum d^2 = 65$

Using the sum of the difference ($\sum d$, see bottom of d column), calculate the mean difference:

$$\bar{d} = \frac{\sum d}{n} = \frac{17}{10} = 1.7$$

Calculate the standard error of the mean difference:

$$SE_d = \sqrt{\frac{\sum d^2 - \frac{(\sum d)^2}{n}}{n(n-1)}} = \sqrt{\frac{65 - (289/10)}{90}}$$

$$= \sqrt{0.4011} = 0.633\,33$$

Calculate t:

$$t = \frac{\bar{d}}{SE_d} = \frac{1.7}{0.633\,33} = 2.68$$

Consult the t table (Table D.2, Appendix D) at df = 9 (number of data pairs minus 1) and $P = 0.05$. The table value is 2.262. Since our value is larger than this, we reject the null hypothesis and, from an examination of the sign of the calculated value of t, can state that there are significantly more invertebrate species above the outfall than below ($t = 2.68$, df = 9, $P < 0.05$).

then the first sample is the larger. Since the sign of *t* is purely arbitrary, we just ignore the sign when examining the table and quoting the results. In this case:

> Significantly fewer species of invertebrates are found below sewage outflows compared to above (paired *t* test: $t = 2.68$, df = 9, $P < 0.05$).

Note that the data in Worked Example 4.3 are listed in two separate columns, with each row representing a single river. If you use a computer program to calculate your paired *t* tests, this format will usually be appropriate. See Appendix B (Table B.2) for details of data entry for statistical analysis using computers.

Wilcoxon matched-pairs test

If we have matched data points, but the data do not lend themselves to parametric analysis (i.e. the data are ordinal or the differences are not normally distributed), then we need a nonparametric equivalent of the paired *t* test. The Wilcoxon matched-pairs test simply involves calculating the differences between pairs of data points and ranking the differences in order of magnitude from small differences to large ones (ignoring the sign of the difference and any values of zero difference). As in all rank tests, tied ranks are given the mean rank (see the procedure outlined in Box 4.4). Next, the sign of the difference is attached to the rank and the positive and negative values are summed separately (and denoted T_+ and T_- respectively).

The assumption is that, if there is no significant difference between the two samples, then the positive and negative ranks should balance. The way we test this null hypothesis is to compare the lowest of the rank sums to a value in a table of critical values (an extract of which is shown in Table 4.4) and if it is equal to or less than the critical value, then we can reject the null hypothesis. If there are cases of zero difference, then the sample size (*n*) used when consulting the table is the number of individuals minus the number of zero differences.

To take an example, we might want to survey a group of people to find whether their attitudes towards the use of taxation to discourage environmentally damaging behaviours are influenced by knowledge of the subject area. We could ask a series of questions and score each person on their attitudes to environmental taxes on a scale

from 1 to 5 (with a low score implying antipathy and a high score indicating commitment). We could then show them an educational video putting the case for and against environmental taxation and question them again. The data obtained from ten respondents could be as in Worked Example 4.4. We have one subject (B) who does not change in score (i.e. the difference in score is zero). This subject is ignored for the rest of the test, so $n = 9$. Five subjects differ by 1, and so get a rank of 3 (the mean rank for the first five positions). Note that the ranks are allocated irrespective of the

WORKED EXAMPLE 4.4 *Calculating a Wilcoxon matched-pairs test statistic for attitudes to environmental taxation before and after watching an educational video*

Respondent	Score before	after	Differences (score after minus score before)	Rank of the differences (ignoring any zeros and the sign)	Signed ranks
A	2	1	−1	3	−3
B	5	5	0	–	–
C	4	5	+1	3	+3
D	2	4	+2	7	+7
E	3	5	+2	7	+7
F	4	5	+1	3	+3
G	3	4	+1	3	+3
H	2	5	+3	9	+9
I	3	4	+1	3	+3
J	2	4	+2	7	+7

From these data we can find the absolute values of the sum of the positive ranks (T_+) and the sum of the negative ranks (T_-):

$$T_+ = 3 + 7 + 7 + 3 + 3 + 9 + 3 + 7 = 42$$
$$T_- = 3$$

The lowest value (irrespective of sign) is 3 so it is this which is compared with the values in Table 4.4.

It is possible to check that T_+ and T_- have been correctly calculated since:

$$T_- = \frac{n(n+1)}{2} - T_+ = \frac{9 \times (9+1)}{2} - 42 = 45 - 42 = 3$$

and:

$$T_+ = \frac{n(n+1)}{2} - T_- = \frac{9 \times (9+1)}{2} - 3 = 45 - 3 = 42$$

Because differences of zero are ignored in this test, ranking begins at the lowest difference greater than 0. If zeros occur, then n is the number of items with differences of more than zero.

Table 4.4 *Selected values of* T *for the Wilcoxon matched-pairs test. Reject the null hypothesis if the calculated* T *value is less than or equal to the table value. Shaded lines indicate the critical values for the example referred to in the text. A more comprehensive table of* T *values is given in Table D.4 (Appendix D)*

n (number of items minus any zero differences)	T	
	P = 0.05	P = 0.01
7	2	–
8	3	0
9	5	1
10	8	3

direction of deviation between the scores (i.e. ignoring the sign of the difference). It is only once the ranks have been allocated that the direction of the deviation is taken into account by attaching the sign of the difference (producing what are called the signed ranks). We can then sum the negative ranks and positive ranks separately. From this example we find that the sum of the negative ranks is 3 and the sum of the positive ranks is 42. The lower of the T values is compared to an appropriate table of critical values (see the shaded area of Table 4.4) at $n = 9$ (where n is the sample size minus the number of zero differences). We find that the lower T value (3) is smaller than the critical value (5) and the probability is therefore less than 0.05. We reject the null hypothesis and say:

> Subjects become more committed to environmental taxation after watching the video ($T = 3$, $n = 9$, $P < 0.05$).

Note that the data in Worked Example 4.4 are listed in two separate columns with each row representing an individual. If you use a computer program to calculate your Wilcoxon matched-pairs tests, this format will usually be appropriate. See Appendix B (Table B.2) for details of data entry for statistical analysis using computers.

As in the Mann–Whitney U test, the test statistic (T) approximates to a z distribution when sample sizes are large ($n > 15$). In addition, there is a correction factor to increase accuracy when there are tied ranks (see Siegel and Castellan, 1988). If your computer program gives you a z value, then record this as well as T and P. Similarly, if the program calculates a correction for ties, record the corrected z and P as well as T. If you are performing calculations by hand, then this inaccuracy is a problem only if you have a marginally non-significant result (P just greater than 0.05). If the result is significant, or P is much greater than 0.05, it will remain so even if corrected for ties.

Summary

By now you should be able to find whether there is a significant difference between:

● two means using a *t* test, and know when this test is appropriate (for comparing two independent samples of normally distributed data);

● two medians using a Mann–Whitney *U* test, and know when this test is appropriate (for comparing two independent samples of ordinal or non-normal data);

- matched pairs of data using a paired *t* test, and know when this test is appropriate (when the differences between the data pairs are normally distributed);

- matched pairs of data using a Wilcoxon matched-pairs test, and know when this test is appropriate (for matched pairs of ordinal data, or when the differences between data pairs are not normally distributed).

Questions

4.1 Two sources of sediment in rivers are construction sites (which are likely to be urban in origin) and agricultural land (rural in origin). To investigate whether there was a difference in the amount of suspended sediments in urban and rural stretches of rivers, a river section in each of ten urban and rural areas was selected (each river was only used once, i.e. the data are independent). The river sections were otherwise similar in size, flow rate, altitude etc. The following sediment loads (in milligrams per litre) were obtained:

Rural	82	55	50	34	50	44	29	77	60	71
Urban	95	111	39	78	56	62	82	93	77	62

(i) The above data have been examined and appear to be normally distributed. Use a *t* test on the data, and record the following:

	Rural	*Urban*
Mean suspended sediment		
Standard deviation		
t value		
df		
Probability		

(ii) What can you conclude about the suspended sediment levels in rural and urban rivers?

4.2 Because persecuting badgers is an illegal activity, badger sets that are more secluded from the view of the general public may be more at risk from persecution. To test this hypothesis, badger sets were ranked as to their visibility to the nearest public right of way (footpath or road), with a score of 1 given to those which it would be impossible for people to see, and a score of 6 indicating that the set was certain to be seen. The sets were then identified as being active (i.e. with a badger social group using the set regularly) or no longer in use (where there were no recent signs of badger activity).

The visibility scores for the two set types are given below:

Active	1	2	4	3	2	2	1	3	5	4	5	4	4	3	5
Unused	2	3	1	2	1	2	3	3	2	4	2	2	2	2	1

(i) Use a Mann–Whitney U test (because the data are ordinal) on the above data, and record the following:

	Active	Not in use
Median score		
Sample size		
Smaller U value		
Probability		

(ii) What do you conclude about the effect of set visibility on whether or not a set is still active?

4.3 In a study of copper contamination of soils in woods near a smelter, the humus layer and the soil layer of each of six woodlands (A to F) were analysed for copper, and the following concentrations of copper were obtained (in micrograms per gram):

Woodland	A	B	C	D	E	F
Humus	155	48	25	24	22	15
Soil	150	24	19	19	15	14

(i) Use a paired t test on the above data (because the humus and soil samples are matched for each woodland) and report the following:

Mean difference (d)		
t value:	df:	Probability:

(ii) What can you conclude about the amount of copper in the humus layer and the soil layer of these six woods?

4.4 In a telephone survey, the managers of 20 small businesses were questioned on their attitude to improving the environmental nature of their products, and given a score for commitment (based on their statements and current efforts, from 0 indicating no interest, to 8 indicating a high level of interest and positive activity in that direction). Following this survey, the companies were each sent the results of research showing the attitudes that potential customers had towards environmentally friendly products of the type produced by their company. The companies were then each given a follow-up telephone survey, and their enthusiasm towards improving the environmental standard of their products was again assessed. The following data for the two telephone surveys were obtained.

Business	A	B	C	D	E	F	G	H	I	J	K	L	M	N	O	P	Q	R	S	T
Before	1	1	4	8	7	2	2	3	1	3	2	2	4	4	2	1	5	2	8	1
After	2	2	5	8	8	1	1	4	2	4	3	3	5	4	1	2	4	4	8	2

(i) Use a Wilcoxon matched-pairs test on the above data (because the answers to each survey are matched for each company) and report the following:

T:	n:	Probability:

(ii) What can you conclude about the attitudes of the managers of small businesses?

4.5 Consider an investigation into the possible effects on non-smokers of allowing smoking in a shared office environment. Which of the four tests explored in this chapter (*t* test, Mann–Whitney *U* test, paired *t* test or Wilcoxon matched-pairs test) would you employ to identify differences between the central tendency of the samples collected if the investigators had:

(i) Measured the lung capacity (in cubic metres) of the non-smokers before smoking was banned, and on the same staff a month after a ban on smoking had been enforced?
(ii) Measured the lung capacity on a scale of 1 to 5 (where 1 was poor and 5 was excellent) of non-smoking staff in an office where smoking was allowed and similarly on staff in an adjacent office where smoking had been banned for a month?
(iii) Measured the lung capacity on a scale of 1 to 5 (where 1 was poor and 5 was excellent) of the non-smoking office staff before smoking was banned, and on the same staff a month after a ban on smoking had been enforced?
(iv) Measured the lung capacity (in cubic metres) of non-smoking staff in an office where smoking was allowed and similarly of non-smoking staff in an adjacent office where smoking had been banned for a month?

5 Relationships between variables

When, for each individual sample unit, more than one continuous or ranked variable is measured, we may wish to quantify the relationship between the variables: as the magnitude of one variable increases, does the other increase or decrease? This chapter covers:

- **Pearson's product moment correlation coefficient for normally distributed data**
- **Spearman's rank correlation coefficient for data measured at least on an ordinal scale**
- **Regression techniques (simple linear regression)**
- **Consideration of circumstances where there are more than two variables being investigated simultaneously**

In Chapter 4, we were concerned with the case where we had taken two samples and wanted to find whether there was a difference between them. Another useful technique in statistics is used when we have only one sample, but for each individual we have measured two variables and are interested in whether there is a relationship between the two variables. For example, we could sample several colliery spoil sites and for each obtain a measure of the pH of the soil and the number of plant species occurring on the site, obtaining the relationship between the two variables shown in Figure 5.1. In this case, there is a positive relationship between the number of plant species and pH, with the number of species increasing as the pH increases.

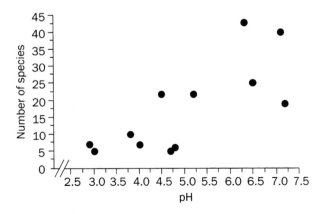

Figure 5.1 *Relationship between the pH of the soil and the number of plant species growing at 13 colliery spoil sites*

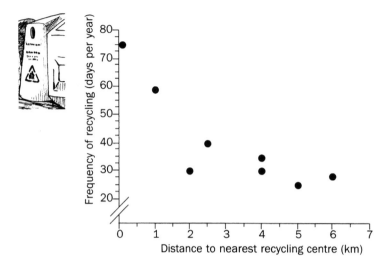

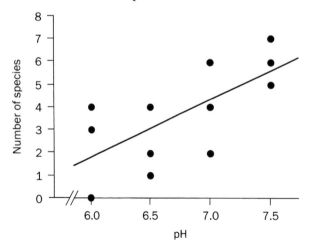

Figure 5.2 *Negative relationship between the frequency of recycling and the distance to the nearest recycling centre (n = 8)*

Two variables may be negatively related. For example, a survey amongst people who recycled their waste might find a negative relationship between the distance they lived from the recycling centre and the frequency with which they visited it (Figure 5.2).

There are two main methods of analysing relationships: **correlation** and **regression**. In correlation, the degree to which two variables vary together is measured. Correlation is, therefore, a measure of the strength of the relationship between two variables. Regression is used to calculate the **line of best fit** through the data, in order to establish or estimate the dependence of one variable upon the other (thus assuming a cause and effect relationship) and/or enable predictions of one variable to be made from knowledge of another. To illustrate the difference, take the colliery spoil data. Here, correlation analysis is appropriate because, although we are interested in how soil pH and number of species vary together, we cannot necessarily assume that pH – rather than one or more related unmeasured variables (such as soil fertility) – causes the observed relationship with number of species. To demonstrate a cause and effect relationship between soil pH

Figure 5.3 *The effect of pH upon the number of plants germinating (n = 12)*

and the number of species, we could perform a controlled experiment, where we took several soils that differ only in pH, and then sowed seeds of several different plant species. We could then use regression analysis to determine whether varying the pH affects the number of species germinating, and might get a relationship such as the one in Figure 5.3.

In the case of regression (unlike correlation), we are justified in fitting a line to our data. The line used is that which best fits the data points. However, it is not fitted by eye: instead we use a mathematical method which we explain later in this chapter. When plotting graphs of regression data, the variable that is controlled by the researcher (in this example, soil pH) is placed along the horizontal axis (also known as the *x* axis, so pH is known as the *x* variable). Along the vertical axis (*y* axis) is the measured variable (here, the number of species germinating; also known as the *y* variable). The measured variable (e.g. the number of species) is the dependent variable, because its value depends on the value of the controlled, or independent, variable (e.g. the soil pH).

When illustrating a correlation, the same logic is used for deciding which variable should be on the *x* and *y* axes. In the relationship between frequency of visits to the recycling centre and the distance travelled to the centre, frequency of visits could be considered to depend on the distance a person has to travel to get to the centre, and therefore is plotted on the *y* (vertical) axis. There would be no logic in considering that the distance to the centre depends on the frequency of visits, therefore distance to the centre is the independent variable and is plotted on the *x* axis. Note that even though we have assigned the variables as being dependent and independent, this is for the purposes of plotting the graph only: in correlation we are not necessarily assuming a causal relationship between the two. In some cases, we cannot assign a pair of variables as dependent and independent. For example, if we had measured two soil characteristics for several soil samples (e.g. conductivity and pH) there would be no reason for assigning either variable as the dependent or independent variable. In such a case, axis allocation is arbitrary.

The first stage in performing correlation or regression analysis is to draw a scatterplot and look at the pattern of the points. This is necessary because the method chosen depends on whether there is a linear relationship between the two variables. A scatterplot alerts us to any non-linear relationship in the data. For example, if the concentration of copper in the soil at various distances from a smelter was measured, we might find a curved relationship between the two variables (Figure 5.4).

In a case like Figure 5.4, we could transform the data (taking the log of the *y* variable would straighten this relationship), or use appropriate methods for non-linear data – see texts such as Zar (1999) or Sokal and Rohlf (1995). Where data appear to show a curved relationship, a variety of transformations (see Chapter 3) can be used on either or both of the *x* or *y* axes to straighten the line (whichever combination gives the best linear relationship) followed by analysis on the transformed data.

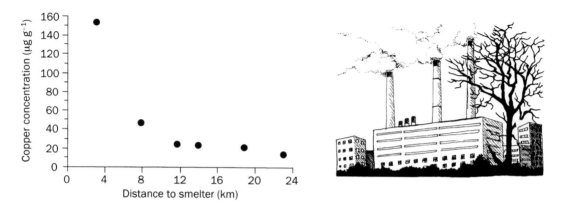

Figure 5.4 *Relationship between distance from a smelter and the copper concentration in the humus layer of the soil of six woods*

Correlations

We use a test statistic called a correlation coefficient to measure the strength of a relationship. Correlation coefficients have values that lie between +1 (perfect positive relationship: Figure 5.5a) and –1 (perfect negative relationship: Figure 5.5b), with 0 being indicative of no relationship at all (Figure 5.5c). Intermediate values imply weaker relationships. When we have a weaker relationship then we need an objective test to decide whether the relationship is significant. Note that a relationship such as that in Figure 5.5d would also have a low correlation coefficient (approaching 0). This is because correlations only measure linear relationships.

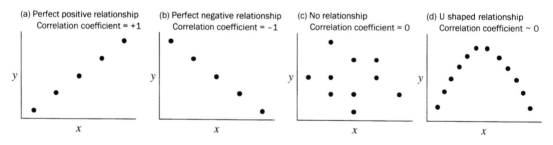

Figure 5.5 *Range of possible correlation coefficients*

The null hypothesis is that there is no relationship. By consulting appropriate statistical tables for any correlation coefficient using the sample size (or degrees of freedom for the parametric correlation: the number of data pairs minus 2) we can find the probability of obtaining any particular correlation coefficient by chance. We compare our calculated correlation coefficient (ignoring the sign at present) at the critical level of $P = 0.05$ to test the null hypothesis of no significant relationship. If the probability of there being no significant relationship is less than 0.05, then we can

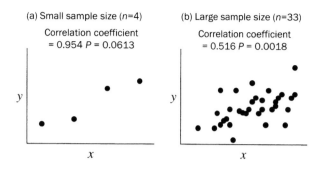

(a) Small sample size (n=4)

Correlation coefficient
= 0.954 P = 0.0613

(b) Large sample size (n=33)

Correlation coefficient
= 0.516 P = 0.0018

Figure 5.6 *Effect of sample size on the significance of a correlation*

reject the null hypothesis and accept an alternative hypothesis that there is a significant relationship. The sign of the correlation coefficient tells us which direction the relationship is in, and values closer to +1 (for positive relationships) or −1 (for negative relationships) indicate stronger correlations. As well as the strength of the relationship (the value of the correlation coefficient), the significance of a correlation depends on the sample size. Figure 5.6a shows that with only four data points, even though the correlation coefficient is almost +1, the correlation is not significant ($P > 0.05$). This is because with only four data points the pattern could easily occur by chance. Figure 5.6b shows a weaker correlation (the correlation coefficient is only around 0.5) which, because of the large sample size (33 data pairs), is statistically highly significant ($P < 0.01$). Therefore, both the size of the correlation coefficient and the sample size influence the resulting P value, as shown in Figure 5.7.

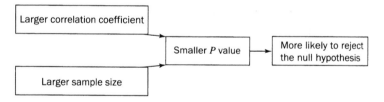

Figure 5.7 *Relationship between the correlation coefficient, sample size and probability*

Remember that significant correlations do not identify cause and effect. Suppose we were interested in any relationship there might be between the incidence of disease in certain areas and the numbers of doctors in those areas. Towns with a large number of reported diseases may also have a large number of doctors. Of course, doctors do not cause the high number of diseases, rather, large towns have a large population and, therefore, a large number of ill people compared to small towns. The number of doctors is likely to be related to the population size rather than directly to the number of disease incidents. Although correlations do not necessarily identify cause and effect, they may indicate relationships between variables, the reasons for which may merit further investigation. In other words, correlations may provide a source of further hypotheses. This example highlights one important problem of correlation analysis called **autocorrelation**, where several variables are correlated and separation of the influence of one from another is difficult. Autocorrelation is a particular problem with climatic variables where wetter days tend to be warmer, have greater cloud cover and fewer sunshine hours. In geography, autocorrelation often occurs with spatial data: individual sample points close to each other in space are likely to be influenced by many similar variables – see, for example, Haining (1993) and Burt and Barber (1996).

There are parametric and nonparametric tests for measuring correlations. We will consider one of each: **Pearson's product moment correlation coefficient** (parametric) and **Spearman's rank correlation coefficient** (nonparametric).

Pearson's product moment correlation coefficient

Having plotted data on a scatterplot, if a vertical line is drawn up from the x axis at the mean value of x, and a horizontal line drawn along from the mean value of y, then the graph would be divided into quadrants (Figure 5.8). Each data point could now be expressed in terms of its position with reference to these lines. Those in quadrant 1 would have negative values in relation to the mean of x and positive values in relation to the mean of y. Multiplying these two values together would give a negative product (because a positive number multiplied by a negative number gives rise to a negative number). If we look at all the quadrants we find that the products from quadrants 1 and 4 are negative, while those from quadrants 2 and 3 are positive (multiplying two negative values in quadrant 3 gives a positive product – see Table 5.1).

We can see from Figure 5.8 that samples which are positively correlated tend to have positive products of deviation from the mean (data points mainly in quadrants 2 and 3) and those which are negatively correlated will have negative products of deviation (data points mainly in quadrants 1 and 4). The sum of the products of deviation (also called the sum of cross products) divided by the number of degrees of freedom (the number of data pairs minus 1) gives us a measure of correlation. Once this has been standardised (by dividing by the standard deviations of x and y) to produce a coefficient on a scale of -1 to $+1$, it is known as Pearson's product moment correlation coefficient (r). A rearranged formula is given in Box 5.1.

The probability of obtaining the computed r value by chance may now be obtained from tables of critical values (an extract of which is shown in Table 5.2) using degrees of freedom calculated as the number of data pairs minus 2. With very large sample sizes, it is possible to obtain significant relationships even when r is comparatively low and the scatterplot

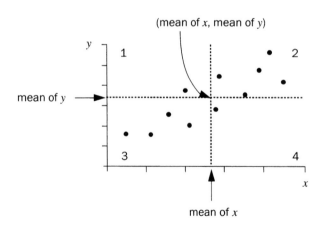

Figure 5.8 *Scatterplot divided into quadrants defined by the mean values of* x *and* y

Table 5.1 *Deviations from mean axis values in scatterplots*

Quadrant	Deviation from the mean of x (i.e. $x - \bar{x}$)	Deviation from the mean of y (i.e. $y - \bar{y}$)	Product of the deviations $(x - \bar{x})(y - \bar{y})$
1	Negative	Positive	Negative
2	Positive	Positive	Positive
3	Negative	Negative	Positive
4	Positive	Negative	Negative

Box 5.1 *Formula for Pearson's product moment correlation coefficient (r)*

The sum of cross products (derived from Table 5.1) is calculated as

$$\sum (x - \bar{x})(y - \bar{y})$$

where:

x and y are the data points on the respective axes;

$\bar{x}$ and $\bar{y}$ are the means of x and y, respectively.

Pearson's product moment correlation coefficient (r) is calculated as the sum of the cross products, divided by the standard deviation of x and the standard deviation of y; this rearranges to give

$$r = \frac{n\sum xy - \sum x \sum y}{\sqrt{\left[n\sum x^2 - \left(\sum x\right)^2\right]\left[n\sum y^2 - \left(\sum y\right)^2\right]}}$$

where:

n is the number of data pairs;

$\sum x$ and $\sum y$ are the sums of x and y, respectively;

$\sum xy$ is the sum of the products of x and y (i.e. each value of x multiplied by its associated value of y and then all summed).

---

shows a high degree of scatter (as previously illustrated in Figure 5.6b). A useful way of determining the importance of the correlation is the **coefficient of determination** (see Box 5.2) which is a measure of the proportion of the variation in one variable that is explained by the variation in the other variable.

The data for the number of plant species and the soil pH of the 13 colliery spoil sites are given in the first two columns of Worked Example 5.1. The results show that $r = 0.700$, which at df = 11 can be compared to the critical values in an appropriate table (see the shaded area of Table 5.2) to obtain $P < 0.01$. By squaring the correlation coefficient, we find that the coefficient of determination (r^2) is 0.490, or $R^2 = 49\%$,

WORKED EXAMPLE 5.1 *Calculating Pearson's product moment correlation coefficient between soil pH and the number of species of plant growing on colliery spoil*

pH	No. of species			
(x)	(y)	x^2	y^2	xy
2.8	17	7.84	289	47.6
2.9	7	8.41	49	20.3
3.8	10	14.44	100	38.0
4.5	22	20.25	484	99.0
7.1	40	50.41	1600	284.0
6.5	25	42.25	625	162.5
3.0	5	9.00	25	15.0
4.7	5	22.09	25	23.5
5.2	22	27.04	484	114.4
4.0	7	16.00	49	28.0
4.8	6	23.04	36	28.8
6.3	43	39.69	1849	270.9
7.2	19	51.84	361	136.8

$\sum x = 62.8$ $\sum y = 228$ $\sum x^2 = 332.3$ $\sum y^2 = 5976$ $\sum xy = 1268.8$

$(\sum x)^2 = 3943.84$ $(\sum y)^2 = 51\,984$

Using the totals at the bottom of the columns:

$$r = \frac{n\sum xy - \sum x \sum y}{\sqrt{\left[n\sum x^2 - (\sum x)^2\right]\left[n\sum y^2 - (\sum y)^2\right]}}$$

$$= \frac{13 \times 1268.8 - 62.8 \times 228}{\sqrt{[13 \times 332.3 - 3943.84][13 \times 5976 - 51\,984]}}$$

$$= \frac{2176}{\sqrt{376.06 \times 25\,704}} = 0.700$$

Consult the table of *r* values at df = *n* − 2 = 11. Our value of 0.700 is larger than the *P* = 0.01 table value (0.684). Therefore there is a highly significant positive correlation between pH and number of species (*r* = 0.700, df = 11, *P* < 0.01).

Since *r* = 0.700, the coefficient of determination (r^2) = 0.7^2 = 0.490 and R^2 = 49%.

Box 5.2 *The coefficient of determination*

If the Pearson's product moment correlation coefficient (r) is +1 or –1, then the variation in both variables is perfectly matched. The nearer r is to zero, the less correlation there is between the variables. The coefficient of determination describes the proportion of the variation that the two variables have in common. Its calculation is very simple, being simply the square of the value of r value (i.e. r^2). Multiplying this value by 100 expresses it as a percentage (called R^2).

Table 5.2 *Selected values of r for Pearson's product moment correlation coefficient. Reject the null hypothesis if the calculated value of r is greater than the table value. Shading indicates the critical values for the example referred to in the text. A more comprehensive table of r values is given in Table D.5 (Appendix D)*

df	r	
$(n-2)$	**P = 0.05**	P = 0.01
8	**0.632**	0.765
9	**0.602**	0.735
10	**0.576**	0.708
11	0.553	0.684
12	**0.532**	0.661

which implies that the number of species and soil pH on the colliery spoil sites have 49% of their variation in common. This does not imply that soil pH directly influences the number of species. Other variables are also likely to be involved; even if the sites were carefully selected to be as similar as possible, they could differ in aspect, nutritional status of the soil, moisture availability, etc. Thus we can say that:

There is a highly significant positive relationship between soil pH and the number of species of plants growing on colliery spoil ($r = 0.700$, df = 11, $P < 0.01$), with the two variables having 49% of their variation in common.

Note that the data in Worked Example 5.1 are listed in two separate columns (one for each variable) with each row representing a single soil sample. This data format is likely to be appropriate for analysis on computer (see Table B.3, Appendix B).

Spearman's rank correlation coefficient

If one or both of the variables are measured on an ordinal scale, or if the data are non-normal, then a nonparametric correlation is required. This technique is also useful if you wish to examine a relationship between two variables which may not be linear, but which none the less shows an increase (or decrease) of one variable with an increase in the other. One of the most commonly used nonparametric correlation coefficients is Spearman's rank correlation coefficient (r_s). This is calculated by ranking each variable separately and comparing the ranks of each data pair.

> ## Box 5.3 Formula for Spearman's rank correlation coefficient (r_S)
>
> $$r_s = 1 - \frac{6\sum d^2}{n^3 - n}$$
>
> where
>
> d is the difference between the ranks within each pair of data points;
>
> n is the number of data pairs.

Suppose we know that seven fields have been left fallow for a different number of years, but we are only sure of the actual number of years for six of these, knowing only that the seventh is very much older than the rest, we could obtain the data below:

Time fallow (years)	1	2	3	4	8	10	>10
Number of plant species (m⁻²)	2	3	5	4	7	6	7

If we wish to examine the relationship between the reduced management and the floristic diversity, we can rank the fields in order of the number of years laid fallow and also in terms of the number of plant species present. Although these data are not suitable for a parametric analysis because we do not know the actual number of years left fallow for the seventh field, they can be examined using a nonparametric test. We first rank the data in each column separately (giving a mean value to any ties: see Box 4.4), and calculate the difference between each pair of ranks. The calculation for Spearman's rank correlation coefficient (r_S) is given in Box 5.3. If the calculated value of r_S is positive, then there is a positive relationship between the two variables. Conversely, if r_S is negative there is a negative relationship. To find if the relationship is significant, we consult a table of critical values of r_S (an extract of which is shown in Table 5.3), ignoring the sign of r_S.

From the data in Worked Example 5.2, we find that $r_S = 0.902$. Examining the shaded area of Table 5.3, we find that with $n = 7$ the probability is less than 0.05 that there is no significant correlation and, since the r_S value is positive, we can say that:

Table 5.3 *Selected values of r_S for Spearman's rank correlation coefficient. Reject the null hypothesis if the calculated value of r_S is greater than the table value. Shading indicates the critical values for the example referred to in the text. A more comprehensive table of r_S values is given in Table D.6 (Appendix D)*

n	r_S	
	P = 0.05	P = 0.01
6	0.886	–
7	0.786	0.929
8	0.738	0.881
9	0.683	0.833
10	0.648	0.794

WORKED EXAMPLE 5.2 *Calculating Spearman's rank correlation coefficient between the floristic diversity and the time fields have lain fallow*

Actual data		Rank of data		Difference between ranks (d)	d^2	Calculation
Time fallow (years)	No. of plant species (m⁻²)	Time fallow	No. of plant species	(rank time fallow – rank no. of plant species)		
1	2	1	1	0	0	$r_s = 1 - \dfrac{6\sum d^2}{n^3 - n}$
2	3	2	2	0	0	
3	5	3	4	−1	1	
4	4	4	3	1	1	$r_s = 1 - \dfrac{6 \times 5.5}{343 - 7}$
8	7	5	6.5	−1.5	2.25	
10	6	6	5	1	1	
>10	7	7	6.5	0.5	0.25	$r_s = 0.902$
					$\sum d^2 = 5.5$	

Since $n = 7$, from Table 5.3, $P < 0.05$

There is a significant positive correlation between the time laid fallow and the number of plant species per square metre ($r_s = 0.90$, $n = 7$, $P < 0.05$).

Usually we would display this type of relationship using a scatterplot. However, in this example we would have difficulty in deciding where to place the final x value (>10) since it is of unknown magnitude compared to the other x values. If a graphical display is required for an example such as this, you could plot the x values as ranks (using the rank values of the third column in Worked Example 5.2), with the axis clearly labelled 'Rank of time lain fallow'.

Note that the data in Worked Example 5.2 are listed in two separate columns (one for each variable) with each row representing a single field. This data format is likely to be appropriate for analysis on computer (see Table B.3, Appendix B).

Note that the formula in Box 5.3 assumes that there are no tied ranks. This simplified formula arises from the fact that when there are no ties, all the ranked values are known (although their order is not). The more tied ranks there are, the less accurate this formula is. With a large numbers of ties, either use a correction – see texts such as Zar (1999) or Siegel and Castellan (1988) – or rank the data as before and use Pearson's product moment correlation on these ranked data. If you are performing calculations by hand, then this inaccuracy is a problem only if you have a marginally significant result (i.e. P is just less than 0.05). If the result is not significant, or highly significant, it will remain so even if corrected for ties. Where a computer output gives correlation coefficients and probabilities corrected for ties, it is these corrected values that should be recorded.

Regression

When we have good reason to suspect that a linear and causal relationship exists between two variables and/or we wish to predict the values of one variable from another, then a regression analysis can be used to find an equation to describe the relationship. For example, if we were to measure the sound level of traffic at different distances from a road, we might expect that the further away from the road we were, the quieter it would become. We could take the measurements and fit a line to the data, and use this to predict the traffic noise at a given distance from the road. If a scatterplot of such data appears to follow a linear relationship then we can determine mathematically the line which best fits these data (called the regression line). If the data appear to show a curved relationship then we can transform the data (see Chapter 3) on one or both axes (whichever combination gives the best linear relationship) and analyse the transformed data (or use an appropriate method for analysing non-linear data – see Sokal and Rohlf, 1995). Remember that the data should be plotted with the independent variable on the x axis (e.g. distance from the road) and the dependent variable on the y axis (e.g. noise level).

Any straight line drawn on a graph with x and y coordinates is described by a simple formula (see Box 5.4). The technique of regression involves mathematically fitting a straight line through a cloud of data points, which will then allow us to predict y from any particular x value. There are several possible techniques to calculate the line, and the one described here (**simple linear regression**) is one of the most commonly used.

If all the points in a scatterplot were exactly in a straight line we could simply use a ruler and join the points. However, the relationships we study in environmental subjects are rarely that straightforward. For example, Figure 5.9 shows the line of best fit (the calculation of which is shown below) for the data on noise levels; this shows that, although there is an obvious negative relationship between noise levels and distance from a road (i.e. the line slopes down and the gradient is negative), there is still some scatter around the line of best fit. The technique of simple linear regression involves finding values of a (the intercept) and b (the gradient) which describe the line of best fit through a cloud of data points. This technique assumes that the x (independent) variable is known without error or is under the control of the

Box 5.4 *The formula for a straight line*

A straight line can be described by the equation

$y = a + bx$

where:

x and y are the coordinates (see below);

a is the intercept of the line on the y axis (i.e. the value of y when $x = 0$);

b is the gradient (slope).

(You may be more familiar with this equation in the form $y = mx + c$.) For any given line, regardless of the value of x and y, the values of a and b are constant. This formula can be used to predict values of y for any value of x if the gradient and intercept are known. In the simple example below, the equation is

$y = 1 + 2x$

The intercept (a, where the line crosses the y axis) is 1. The gradient (b, slope) of the line is 2, therefore for every one unit along the x axis, y increases by 2 units. Using the formula above, when x is 1,

$y = 1 + 2 \times 1 = 3$

and when x is 2,

$y = 1 + 2 \times 2 = 5$

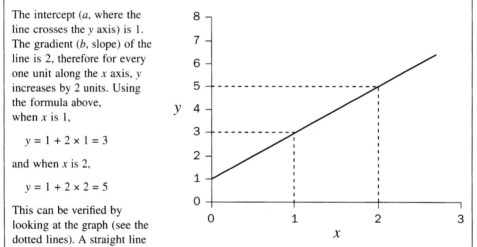

This can be verified by looking at the graph (see the dotted lines). A straight line is uniquely identified by knowing a and b. Points along the line are known as coordinates. Thus, the point where x is 1 and y is 3, has the coordinates (1, 3).

If b is negative, then the line slopes the other way (e.g. for the line $y = 1 - 2x$, for every unit along the x axis, y decreases by 2 units).

investigator (in this case this assumption holds because the distances away from the road have been selected: for other implications of this, see the assumptions of simple linear regression later in this chapter). The y variable can vary for each value of x, so if at 100 m away from the road, several y readings had been taken we would expect the y values to be normally distributed.

There are two major steps: first fitting the line and then calculating the significance of the regression. Since here we want to determine how much of the variation in y is determined by x, this technique is sometimes called the regression of y on x.

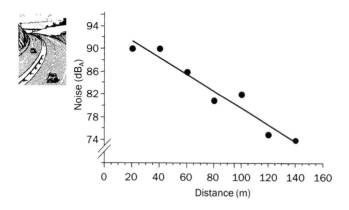

Figure 5.9 *Relationship between noise level and distance from a road (n = 7)*

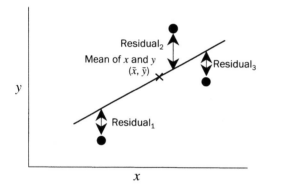

Figure 5.10 *Relationship between the line of best fit and the residuals*

The line of best fit is calculated so that it passes through the mean of x and y (i.e. the coordinate $(\bar{x}, \bar{y})$, and so that the vertical distance from each point to the line is as small as possible. The distance of each point from the line is called the **residual** distance. As can be seen in Figure 5.10, moving the line will decrease the residual distance of some points, but increase it for other points. An infinite number of lines could be drawn on the graph, all pivoting around the coordinate $(\bar{x}, \bar{y})$, but the line of best fit is the one which has the lowest total squared residuals. The technique for finding the line with the lowest total squared residuals is called least squares; squaring the residuals places greater weight on points further from the line and makes all the values positive.

The gradient (b) of the regression line is obtained by dividing the sum of cross products (mentioned in the section on Pearson's product moment correlation coefficients: Box 5.1) by the **sum of squares** of x (first introduced in Box 2.3). The intercept (a) can then be calculated by substituting the value of b together with the mean values of x and y in the formula for a straight line. The formulae for calculating a and b are given in Box 5.5.

The gradient (b) is also known as the regression coefficient. It is worth pointing out at this stage that the formula for calculating b has some similarities with the one used earlier to calculate r (Pearson's product moment correlation coefficient: Box 5.1). In regression, the value of r itself is meaningless, and should not be quoted in the results section of a project report. This is because it assumes that both the x and y values are randomly distributed – an assumption that is not met in regression since the x values are under the control of the investigator. However, the coefficient of determination (r^2 or, as a percentage, R^2) is useful in identifying the proportion of the total variation in y that is explained by the variation in x. Although this can be calculated using the square of the correlation coefficient (Box 5.2), once all of the components required for regression analysis have been calculated, r^2 may be calculated directly in a simpler

---

Box 5.5 *Formulae for the gradient and intercept of the regression line*

The gradient (*b*) can be calculated as the sum of cross products divided by the sum of squares of *x*, which rearranges to give

$$b = \frac{\sum xy - \dfrac{\sum x \sum y}{n}}{\sum x^2 - \dfrac{\left(\sum x\right)^2}{n}}$$

where:

n is the number of data pairs;

$\sum x$ and $\sum y$ are the sums of *x* and *y*, respectively;

$\sum xy$ is the sum of the products of *x* and *y* (i.e. each value of *x* multiplied by its associated value of *y* and then all summed).

Since the line passes through the point described by the two means ($\bar{x}$, $\bar{y}$), the intercept (*a*) is calculated by rearranging the formula for a straight line ($y = a + bx$) to give

$$a = \bar{y} - b\bar{x}$$

---

manner (see later, Box 5.7). In Worked Example 5.3a, both the gradient and intercept are calculated from our data for the relationship between noise and the distance from the road. We get a negative number for *b* of –0.146. This tells us that there is a negative relationship between *x* and *y*: that is, *y* decreases by 0.146 for every unit increase in *x*. Then, once the intercept (*a*) has been calculated as 94.29, the formula for the line (illustrated in Figure 5.9) becomes $y = 94.29 - 0.146x$.

To fit the regression line to the scatterplot, we need to calculate two coordinates and then join them. We already know the values of two of these coordinates: the intercept (0, *a*) and the means of *x* and *y* ($\bar{x}$, $\bar{y}$). However, when plotting a graph by hand it is more accurate to have the points further apart. In Worked Example 5.3a we take the values of *x* of 20 and 140 and calculate that the coordinates are (20, 91.37) and (140, 73.85). These can then be plotted on the scatterplot and the points joined to produce the line shown in Figure 5.9. The line drawn should not exceed the limits imposed by the measured data. This is because we have no information regarding the shape of the relationship at distances below 20 m or above 140 m. In fact, we might expect the line to stop dropping eventually and to flatten out once the distance from the road becomes sufficient that traffic noise is no longer an issue. Note the implications of this: we are assuming that the relationship is linear only within the limits of our data.

Note that the data in Worked Example 5.3a are listed in two separate columns (one for each variable) with each row representing a single noise sample. This data format is likely to be appropriate for analysis on computer programs (see Table B.3, Appendix B).

WORKED EXAMPLE 5.3a *Regression of traffic noise level (y) on distance from road (x): calculation of the equation of the regression line*

Distance (m)		Noise (dB$_A$)		
x	x^2	y	y^2	xy
20	400	90	8 100	1 800
40	1 600	90	8 100	3 600
60	3 600	86	7 396	5 160
80	6 400	81	6 561	6 480
100	10 000	82	6 724	8 200
120	14 400	75	5 625	9 000
140	19 600	74	5 476	10 360
$\sum x = 560$	$\sum x^2 = 56\ 000$	$\sum y = 578$	$\sum y^2 = 47\ 982$	$\sum xy = 44\ 600$

$$b = \frac{\sum xy - \frac{\sum x \sum y}{n}}{\sum x^2 - \frac{(\sum x)^2}{n}} = \frac{44\ 600 - \frac{560 \times 578}{7}}{56\ 000 - \frac{560^2}{7}} = \frac{-1\ 640}{11\ 200} = -0.146$$

$$a = \bar{y} - b\bar{x} = \frac{578}{7} - \left(-0.146 \times \frac{560}{7}\right) = 94.29$$

The regression equation is thus

$$y = 94.29 - 0.146x$$

(the negative sign of b tells us that the slope is negative).

To fit the regression line by hand we take one low value of x (20) and one high value (140) and calculate the value of y for each. When $x = 20$,

$$y = 94.29 - 0.146 \times 20 = 91.37$$

When $x = 140$,

$$y = 94.29 - 0.146 \times 140 = 73.85$$

These two points can then be plotted and the line drawn between them.

Testing the significance of the regression line

We now need to know whether the calculated line is a significant one – in other words, whether there is a significant relationship between traffic noise and the distance from the road. We can do this using a technique known as **analysis of variance** (abbreviated to **ANOVA**). This is the most commonly used method, although a one-sample *t* test could also be employed to determine whether the gradient is significantly greater than zero – see texts such as Zar (1999) or Sokal and Rohlf (1995) for further details. In calculating the significance of the regression, we work out the total variation in *y* and find out how much of this total variation can be explained by variation in *x*. The total variation in *y*, also called the total sum of squares (SS_{total}), is calculated by subtracting the mean of *y* from each value of *y*, squaring the answer, and then adding these squared deviations together. The deviations from the mean value of *y* are illustrated in Figure 5.11, and a rearrangement of the formula is shown next to the figure (you may recognise this as the first stage in calculating the standard deviation in Chapter 2: see Box 2.3).

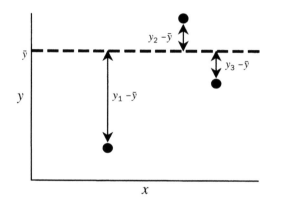

$$SS_{total} = \sum y^2 - \frac{\left(\sum y\right)^2}{n}$$

Figure 5.11 Illustration of the deviations of y from the mean of y

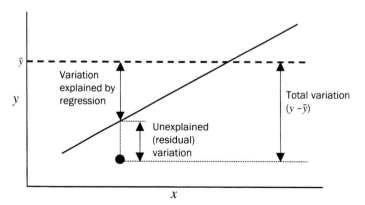

Figure 5.12 The components of variation for regression (as illustrated by a single data point)

This variation in *y* (total sum of squares of *y*) can be broken down into two parts, the variation that is explained by the regression line ($SS_{regression}$) and the unexplained variation (the residual variation or $SS_{residual}$). In other words, $SS_{total} = SS_{regression} + SS_{residual}$. These different components of the variation are illustrated for a single data point in Figure 5.12.

Box 5.6 *Formula for the sums of squares for a regression analysis*

First calculate the total variation in y (SS_{total}), i.e. the sum of squares of y (illustrated in Figure 5.11):

$$SS_{total} = \sum y^2 - \frac{\left(\sum y\right)^2}{n}$$

Then calculate the variation due to the regression (explained) and the variation due to the residuals (unexplained).

The variation in y which is explained by x (regression sum of squares) is the sum of the squares of the vertical distances between the expected values of y (i.e. the regression line) and the mean value of y. This is calculated as the sum of the cross products squared (which we first came across in Box 5.1), which is divided by the sum of squares of x:

$$SS_{regression} = \frac{\left(\sum xy - \frac{\sum x \sum y}{n}\right)^2}{\sum x^2 - \frac{\left(\sum x\right)^2}{n}}$$

The variation in y which cannot be explained by x (residual sum of squares) is the sum of squares of the vertical distances between the measured values of y (the points) and the expected values of y (the regression line). The residuals are also called errors (i.e. what remains of the variation in y when the effect of x has been removed). The residual sum of squares is calculated by subtracting the regression sum of squares from the total sum of squares:

$$SS_{residual} = SS_{total} - SS_{regression}$$

Table 5.4 *Producing an ANOVA results table*

	df [a]	Sums of squares (SS)	Mean squares (MS)	F value	P value
Regression	1	$SS_{regression}$	$\dfrac{SS_{regression}}{1}$	$\dfrac{MS_{regression}}{MS_{residual}}$	Look up in F tables (e.g. Table 5.5) at regression degrees of freedom (1) and residual degrees of freedom ($n-2$)
Residual	$n-2$	$SS_{residual}$	$\dfrac{SS_{residual}}{n-2}$		
Total	$n-1$	SS_{total}			

[a] n is number of data pairs.

ANOVA uses the ratio of the explained variation to the unexplained variation. If we had a perfect fit of our data to the regression line, the residual variation would be zero and the total sum of squares of y would equal the regression sum of squares. That is, the regression line would describe all the variation in y. Conversely, if there was no fit whatsoever, the regression sum of squares would be zero (i.e. the regression line would be horizontal) and the total sum of squares of y would equal the residual sum of squares. The formulae for calculating the sums of squares are given in Box 5.6. These sums of squares are then entered into an ANOVA table such as Table 5.4. Here, the sums of squares (total, regression and residual) are listed with their respective degrees of freedom (the rules for calculating these are also given in Table 5.4). The sums of squares are standardised by dividing by their degrees of freedom to produce measures of variance called **mean squares**. The test statistic (F) is simply the ratio of explained to unexplained variation; in this case the ratio between the regression mean square and the residual mean square. This value is then looked up in an F table (an extract of which is shown in Table 5.5) using two values of degrees of freedom: that of the regression and that of the residual.

Table 5.5 *Selected values of F for ANOVA. Reject the null hypothesis if the calculated values of F are greater than the table values. The upper, bold values are for P = 0.05, while the lower figures are for P = 0.01. Shaded lines indicate the critical values for the example referred to in the text. A more comprehensive table of F values is given in Table D.7 (Appendix D)*

		Regression degrees of freedom				
		1	2	3	4	5
	1	**161**	**200**	**216**	**225**	**230**
		4052	5000	5403	5625	5764
	2	**18.5**	**19.0**	**19.2**	**19.2**	**19.3**
		98.5	99.0	99.2	99.2	99.3
	3	**10.1**	**9.55**	**9.28**	**9.12**	**9.01**
		34.1	30.8	29.5	28.7	28.2
Residual degrees of freedom	4	**7.71**	**6.94**	**6.59**	**6.39**	**6.26**
		21.2	18.0	16.7	16.0	15.5
	5	**6.61**	**5.79**	**5.41**	**5.19**	**5.05**
		16.3	13.3	12.1	11.4	11.0
	6	**5.99**	**5.14**	**4.76**	**4.53**	**4.39**
		13.7	10.9	9.78	9.15	8.75
	7	**5.59**	**4.74**	**4.35**	**4.12**	**3.97**
		12.2	9.55	8.45	7.85	7.46

WORKED EXAMPLE 5.3B *Regression of traffic noise level on distance from road: testing the significance of the regression line*

First we calculate the sums of squares. Note that the components used in the calculations below have already been calculated in Worked Example 5.3a. The total sum of squares is:

$$SS_{total} = \sum y^2 - \frac{\left(\sum y\right)^2}{n} = 47\,982 - \frac{578^2}{7} = 255.714\,29$$

The regression sum of squares is calculated as:

$$SS_{regression} = \frac{\left(\sum xy - \frac{\sum x \sum y}{n}\right)^2}{\sum x^2 - \frac{\left(\sum x\right)^2}{n}} = \frac{\left(44\,600 - \frac{560 \times 578}{7}\right)^2}{56\,000 - \frac{560^2}{7}} = \frac{(-1\,460)^2}{11\,200} = 240.142\,86$$

Now subtract the latter from the former to give the residual sum of squares:

$$SS_{residual} = SS_{total} - SS_{regression} = 255.714\,29 - 240.142\,86 = 15.571\,43$$

Substituting these into an ANOVA table, we get:

	df	SS	MS	F	P
Regression	1	240.14	$\frac{SS_{regression}}{1} = 240.14$	$\frac{MS_{regression}}{MS_{residual}} = 77.11$	<0.01
Residual	$n - 2 = 5$	15.57	$\frac{SS_{residual}}{n-2} = 3.11$		
Total	$n - 1 = 6$	255.71			

In Worked Example 5.3b, the test statistic (F) is calculated to be 77.11. We then look up the value of F in a table of critical values (see the shaded area of Table 5.5). Critical values of F have two values for the degrees of freedom. In this case, we look up F at df = 1 and 5 (sometimes written as $F_{1,5}$). The first figure for the degrees of freedom (1) comes from the regression sum of squares. For simple linear regression, the degrees of freedom for the regression sum of squares are always 1. The second figure for the degrees of freedom (5) comes from the residual sum of squares ($n - 2$). Looking at the shaded section of Table 5.5 we find that the critical value at $F_{1,5}$ is 6.61 (at $P = 0.05$). Since our calculated value of F (77.11) is much higher than either this or the value at $P = 0.01$ (16.3), we can reject the null hypothesis and conclude that a highly significant amount of the variation in y is accounted for by the variation in x.

When quoting results in a report, you should also quote the coefficient of determination, r^2 (or R^2 when converted to a percentage), which is the proportion of the total variation in y that is explained by x (see Box 5.7). For our example (see Worked Example 5.3c) we find that the regression equation explains a high proportion of the variation in y ($R^2 = 93.9\%$), and therefore our regression equation would be a good model for predicting the sound level with distance from the road. We could then report our result as follows:

> There is a significant decrease in sound level with distance from the road ($y = 94.29 - 0.146x$, $F_{1,5} = 77.11$, $P < 0.01$). The regression model explains a high percentage of the variation in sound level ($R^2 = 93.9\%$).

Most computer programs use ANOVA to test the significance of the regression line. Some also display the results of a t test. If both sets of results are given, you only need to record the results of the ANOVA. If you are using a program which only calculates a t test, you should record the t value, number of degrees of freedom ($n - 2$), and the probability (P), in addition to the regression equation (the intercept and slope) and R^2. Interpretation of the t test is the same as when using an ANOVA, in that if the probability is less than 0.05, then there is a significant regression of y on x.

Box 5.7 *Formula for the coefficient of determination for regression analysis*

The coefficient of determination, r^2, can be calculated by first calculating r and then squaring it (see Box 5.2). Alternatively, we can employ the following formula using the sums of squares as calculated in Box 5.6:

$$r^2 = \frac{SS_{\text{regression}}}{SS_{\text{total}}}$$

An r^2 value of one would mean that we had explained all of the variation in y using our regression equation (i.e. $SS_{\text{regression}}$ would equal the total SS_{total}). Conversely, an r^2 of close to 0 would mean that very little of the total variation in y could be explained by the regression line. The value can be converted to a percentage by multiplying by 100 to give R^2.

WORKED EXAMPLE 5.3C *Regression of traffic noise level on distance from road: calculating the coefficient of determination*

Using the sums of squares calculated in Worked Example 5.3b, we obtain:

$$r^2 = \frac{SS_{\text{regression}}}{SS_{\text{total}}} = \frac{240.14}{255.71} = 0.939 \; (R^2 = 93.9\%)$$

So the regression equation explains 93.9% of the variation in y.

Using the regression equation as a predictive model

Now that we have calculated the line, and found that it explains a high proportion of the variability in y, we have a useful **predictive model** enabling us to obtain y for any given value of x. If our regression equation had been found to have a low predictive value, then it may have been worth looking for other factors that may influence y and building a more complex equation, by a method known as multiple regression – see texts such as Tabachnick and Fidell (1996) or Sokal and Rohlf (1995). In our example, R^2 is high (93.9%), therefore the model explains a high degree of the variation in y. If we were 65 m from the road, then from the equation in Worked Example 5.3a we would predict that the noise level should be 84.8 dB$_A$:

$$y = 94.29 - 0.146 \times 65 = 84.8 \text{ dB}_A$$

However, because the intercept and gradient are estimates of true population parameters, there is error around the line. We may therefore wish to attach **confidence limits** to this prediction. The formula for the confidence limits is shown in Box 5.8. The calculation of these for our example is shown in Worked Example 5.3d where we find that the noise level at 65 m from the road is predicted to be between 79.91 dB$_A$ and 89.69 dB$_A$. Note that the prediction of y values from the line is only reasonable within the limits of x used to create the equation, in case the shape of the relationship changes at values outside these limits.

Assumptions of simple linear regression

In addition to the assumption that the relationship between x and y is linear, several other assumptions which apply when using regression are listed below:

1 The independent variable (x) is fixed by the researcher or is known without error (i.e. is not a measured variable). In the example used here, the different distances away from the road were decided before the survey. The value of x is therefore known without error. For many environmental research projects, both x and y are measured variables, but we may still wish to make predictions of one from the other. It is suggested that, even if there is error in measuring x, providing the magnitude of that error does not increase or decrease with the magnitude of x, then simple linear regression is still a valid technique – see, for example, Sokal and Rohlf (1995). Checking the residuals (see 3 below) will help to identify whether such an effect is taking place. Where this assumption does not hold, there are alternative techniques available (called **model II techniques** to distinguish them from simple linear regression, which is a **model I technique**). However, although model II regression methods will calculate the regression line, they should not be used for predictions – see Sokal and Rohlf (1995) for further details.

2 For each known value of x, the value of the dependent variable (y) is free to vary, and should have a normal distribution around each value of x. Thus we are assuming that, for a given value of x metres along the road, if we had obtained

Box 5.8 Formula for confidence limits of a prediction from the regression line

We can calculate the estimated value of y by substituting the value of x at which we wish to estimate y into the regression equation,

$y = a + bx'$

where: x' is the value of x for which we wish to estimate y.

Next, we need to calculate the standard error of the prediction.

$$SE \text{ of prediction} = \sqrt{\frac{SS_{residual}}{n-2} \times \left[1 + \frac{1}{n} + \frac{(x' - \bar{x})^2}{SS_x}\right]}$$

where:

x' is as before;

n is the number of data pairs;

$SS_{residual}$ is the sum of squares of the residuals (see Box 5.6);

SS_x is the sum of squares of x,

$$SS_x = \sum x^2 - \frac{(\sum x)^2}{n}$$

From Box 2.6 we know that when n < 30, the 95% confidence limits are calculated by:

95% confidence limits = estimate ± ($t \times SE$)

Obtaining t at $n - 2$ degrees of freedom and $P = 0.05$, we can calculate the limits within which we are 95% certain that any reading of y would fall if it were taken at x'.

Note that, because the standard error of the prediction increases with greater deviations of x' from the mean value of x (i.e. $(x' - \bar{x})^2$ gets larger as x' gets further from $\bar{x}$), predictions near to $\bar{x}$ are more accurate.

many sound level readings, then there would be a normal distribution of the sound level for that particular distance. In practice, this assumption is difficult to test. Even if multiple samples are taken for each value of x, we do not usually have enough data to inspect for normality.

3   If multiple readings had been taken for each value of x, the variance would be the same along the regression line (i.e. the variance in y would not increase or decrease as x increases). Although this is not easy to test directly, examination of the residuals is a useful alternative. Each residual (i.e. the vertical distance that each point is away from the regression line) is plotted against the measured values of x. The residuals should not vary systematically with a change in the value of x. For example, we should not find patterns such as those illustrated in Figure 5.13.

WORKED EXAMPLE 5.3D *Regression of traffic noise level on distance from road: predicting a value of* **y** *from the regression line (e.g. when* **x** *= 65 m)*

Using the equation obtained from Worked Example 5.3a, the new value of y is estimated at $x = 65$ m:

$$y = 94.29 - 0.146 \times 65 = 84.8 \ \text{dB}_A$$

Next, calculate the sums of squares of x (note that the components are already calculated in Worked Example 5.3a):

$$SS_x = \sum x^2 - \frac{\left(\sum x\right)^2}{n} = 56\,000 - \frac{560^2}{7} = 11\,200$$

and then, using the value of SS_{residual} calculated earlier (15.57: Worked Example 5.3b), and inserting the value of x' (65) for which we wish to estimate y into the formula from Box 5.8:

$$SE \text{ of prediction} = \sqrt{\frac{SS_{\text{residual}}}{n-2} \times \left[1 + \frac{1}{n} + \frac{\left(x' - \bar{x}\right)^2}{SS_x}\right]}$$

$$= \sqrt{\frac{15.57}{5} \times \left[1 + \frac{1}{7} + \frac{(65-80)^2}{11\,200}\right]} = \sqrt{3.114 \times 1.162\,946\,4} = 1.903\,001\,6$$

Since $n < 30$, the 95% confidence limits are calculated by:

95% confidence limits = estimate $\pm (t \times SE)$

The degrees of freedom are $n - 2$, and the value of t at $P = 0.05$ and df = 5 is 2.571. Our estimated value of y was 84.8 dB_A. Therefore:

95% confidence limits = 84.8 $\pm$ (2.571 $\times$ 1.903 001 6) = 84.8 $\pm$ 4.89 dB_A

So

Upper limit = 89.69 dB_A
Lower limit = 79.91 dB_A.

This means that if we were to take a single reading at 65 m from the road, we would be 95% confident that the sound level lay between 79.91 dB_A and 89.69 dB_A.

In Figure 5.13a there is an increase in unexplained variation with increase in x value. In Figure 5.13b the curved relationship between the residuals and x suggests that a curved relationship may exist between y and x. In either case we should reconsider our use of regression, and if we do continue, the data should be transformed. It is also important to look out for clusters of residuals (which may indicate an autocorrelation with a third, unmeasured variable) and for single unusually large residuals (single

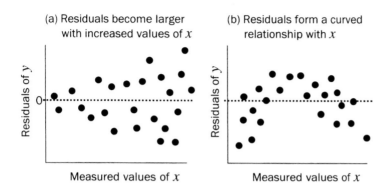

Figure 5.13 *Residuals plotted against the independent variable*

large values can greatly distort the regression line and may make the regression unreliable). To calculate the residuals, we first calculate the predicted values of y for each x value, and then subtract this predicted value from the actual observed values. For our example of sound levels at different distances away from a road, these are calculated in Worked Example 5.3e. These residuals are then plotted on the y axis against either the original x values (as shown in Figure 5.14) or the fitted (predicted) values of y. It does not matter whether you plot the residuals against x or the predicted values of y; if the regression is positive the same pattern will be seen. If the regression is negative, the residual against fitted y value plot will show a mirror image of that seen by plotting the residuals against the x values.

WORKED EXAMPLE 5.3E *Regression of traffic noise level on distance from road: calculating the residuals*

x	Predicted y from the equation obtained from Worked Example 5.3a ($y = 94.29 - 0.146\,x$)	Measured values of y	y residuals (measured y minus predicted y)
20	$y = 94.29 - (0.146 \times 20) = 91.37$	90	−1.37
40	$y = 94.29 - (0.146 \times 40) = 88.45$	90	1.55
60	$y = 94.29 - (0.146 \times 60) = 85.53$	86	0.47
80	$y = 94.29 - (0.146 \times 80) = 82.61$	81	−1.61
100	$y = 94.29 - (0.146 \times 100) = 79.69$	82	2.31
120	$y = 94.29 - (0.146 \times 120) = 76.77$	75	−1.77
140	$y = 94.29 - (0.146 \times 140) = 73.85$	74	0.15

In Figure 5.14 the residuals appear to be randomly positioned with respect to x and therefore our use of simple linear regression is justified. For the purposes of this book our interest in the residuals ends here, but when analysing spatial geographical data, analysis of the residuals themselves may be a useful procedure – see, for example, Haining (1993) and Burt and Barber (1996).

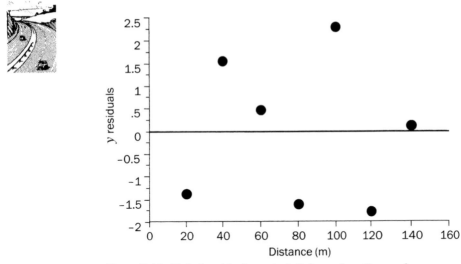

Figure 5.14 *Plot of residuals against distance from the road*

Fitting a regression line through the origin

There are some occasions when we know that the line should go through the origin of the graph (i.e. the point at which x and y equal zero). This is a special case of regression and different formulae apply. For example, we might examine the relationship between the amount of seed damaged by moths and the numbers of moth larvae present in a grain store. It is reasonable that with no larvae (i.e. $x = 0$) we might expect no moth damage (i.e. $y = 0$). The calculations of the slope, sums of squares, and therefore the ANOVA test for significance are also different for this situation – Zar (1999) gives details of these calculations.

Relationships between more than two variables

There are occasions when we are interested in relationships between several variables. Here we briefly describe some of the methods which could be employed, and direct the reader towards further reading for more details.

Correlation matrix

If we simply wished to examine a large number of possible interrelationships (e.g. looking at how climatic variables such as wind speed, cloud cover, rainfall and maximum temperature are related to each other), then we could produce a table (called a correlation matrix) like Table 5.6 with all of the variables across the top and

Table 5.6 *Example of a correlation matrix. The cells contain the Pearson product moment correlation coefficients (r) for each pair of variables. Either parametric or nonparametric coefficients may be used*

	Minimum temperature (°C)	Maximum temperature (°C)	Sun (h)	Rain (mm)	Cloud (oktas[a])	Pressure (mb)	etc.
Maximum temperature (°C)	0.704						
Sun (h)	−0.200	−0.391					
Rain (mm)	0.413	0.225	−0.129				
Cloud (oktas[a])	0.224	0.486	−0.786	0.194			
Pressure (mb)	−0.079	0.213	−0.163	−0.554	0.138		
Wind speed (knots)	0.415	0.220	0.100	0.412	0.017	−0.397	
etc.							

[a] Cloud cover is measured in oktas, a scale from 0 (no cloud cover) to 8 (total cloud cover).

repeated down the side. At each row–column intersection we would place the correlation coefficient between the pair of variables. Such a matrix would not normally be used for significance testing, since, with a large number of variables, several might be significant by chance (e.g. with 5 variables we would have $4 \times 3 \times 2 \times 1 = 24$ tests). Instead, visual examination of patterns in the data would identify pairs of variables which are likely to be related. Either parametric or nonparametric methods of calculating the correlation coefficients can be used in a correlation matrix. More advanced techniques which arise from this include those for reducing a large number of variables to a series of supervariables or factors. These are combinations of the original variables which are correlated within the factor, but not between factors (e.g. factor analysis and principal component analysis – see Tabachnick and Fidell, 1996).

Multiple correlation

If we had several variables which we suspected were correlated, we might be interested in testing the results statistically using multiple correlation, which is similar to correlation analysis but operates on more than two variables. Multiple correlation examines the overall relationship between more than two variables which are normally distributed. The relationship between any two of the variables within the multiple correlation can be examined by a technique called partial correlation, which holds constant the values of the other variables. Texts such as Zar (1999) give details of multiple correlations.

Concordance

There are situations where we might wish to ask whether two or more sets of measurements agree with each other (i.e. they concur). For example, if we had used three different pieces of equipment to measure the conductivity of a series of soil samples, we could be interested in whether the measurements were similar from the different types of equipment. Or, if several observers were rating the aesthetic appeal of several country parks, we might wish to see if the observers agree. The appropriate way of doing this would be to use a technique called concordance correlation. There are parametric (see, for example, Zar, 1999) and nonparametric (e.g. Kendall's coefficient of concordance: Siegel and Castellan, 1988) tests for concordance.

Multiple regression and logistic regression

To obtain a predictive model with several independent variables we would use an extension of regression analysis called multiple regression analysis. There are several options, some of which incorporate all potential independent variables (e.g. soil and climatic variables) into an equation to predict the dependent variable (e.g. crop yield). The simple regression equation of Box 5.4 would be extended to $y = a + bx_1 + cx_2 + dx_3 + \ldots$ (with x_1, x_2 and x_3 being the independent variables, and b, c and d being the relevant slopes). Another method (stepwise multiple regression) builds the model with just those variables which have a major influence, either automatically accepting and rejecting variables in turn, or allowing the researcher to input or extract variables in turn. A feature of multiple regression techniques is that relationships between independent variables are taken into account. This is a parametric technique and, therefore, ordinal variables are inappropriate. However, binary data (i.e. nominal data with two categories, such as male/female, or the presence/absence of a particular species) are permissible. There is another technique (logistic regression) which allows prediction of a nominal scale variable (for example, whether it will rain/not rain within the next week) from a series of continuous, ordinal or nominal variables. Texts such as Tabachnick and Fidell (1996) and Sokal and Rohlf (1995) give details of both multiple and logistic regression.

Summary

By now you should:

● be able to distinguish between a dependent and an independent variable, and know which axis to plot each on (i.e. the dependent variable on the y axis and the independent on the x axis);

● know when to use correlation techniques (to measure the degree to which two variables vary together) and when to use regression techniques (to investigate a causal relationship or to predict values of the dependent variable);

- know that Pearson's product moment correlation coefficient should only be applied when both variables are interval or ratio data, and when inspection of the scatterplot indicates a linear (rather than a curved) relationship (following transformation if necessary);

- know that Spearman's rank correlation coefficient is appropriate if one or both variables are at least on an ordinal (or ranked) scale;

- be able to calculate (and assess the significance of) correlation coefficients and know that a positive value of r (for Pearson's) or r_S (for Spearman's) indicates a positive relationship, a negative value indicates a negative relationship and a value close to zero suggests that there is no linear relationship between the variables;

- be able to calculate the coefficient of determination to assess the proportion (using r^2) or percentage (using R^2) of variation which the variables have in common;

- be able to apply regression techniques (i.e. calculate the gradient and intercept of a slope and test the slope for significance using an analysis of variance), know the assumptions of regression analysis, and be able to inspect the residuals to ensure that their distribution complies with the assumptions of regression (i.e. that they are randomly distributed);

- be aware of other techniques which are appropriate when the relatedness of more than two variables is being examined.

Questions

5.1 For several African countries, the number of physicians (per 10 000 population) and the infant mortality rate (number of deaths of infants less than one year old per 1000 live births) are recorded below:

Country	Number of physicians	Infant mortality rate	Country	Number of physicians	Infant mortality rate
A	0.2	126	J	0.8	116
B	0.5	102	K	0.3	124
C	1.3	65	L	1.3	96
D	0.1	142	M	0.5	117
E	1.4	121	N	0.7	142
F	1.2	64	O	0.5	107
G	1.1	79	P	0.8	85
H	0.4	138	Q	1.4	71
I	0.4	159			

(i) Draw a scatterplot of the data.

(ii) Carry out Pearson's product moment correlation analysis on the data and record the following:

Pearson's r:	df:	P:

(iii) What do you conclude about the relationship between the number of physicians and the infant mortality rate?

5.2 In deciding a policy on admissions charges into a nature reserve, the management conducted a visitor survey on opinions about the level of charges. Two of the questions posed, the choice of responses and the codes for analysis are shown below:

How frequently do you visit the park?

Choice given to respondent	Code
more than once a week	6
once a week	5
once a fortnight	4
once a month	3
several times a year	2
once a year or less	1

What is the maximum you would be prepared to pay to enter this park?

Choice given to respondent	Code
50p or less	1
up to 75p	2
up to £1.00	3
up to £1.25	4
up to £1.50	5
up to £1.75	6
up to £2.00	7
up to £2.50	8
£2.50 or more	9

The results from questioning 20 people were as follows:

Person	*Frequency of visit*	*Amount prepared to pay*	*Person*	*Frequency of visit*	*Amount prepared to pay*
A	3	5	K	4	3
B	3	4	L	2	7
C	2	9	M	1	8
D	4	7	N	3	3
E	2	2	O	4	6
F	1	3	P	2	8
G	6	1	Q	5	1
H	2	9	R	3	6
I	3	5	S	4	5
J	3	3	T	6	2

(i) Carry out Spearman's rank correlation on the data, and record the following:

r_S:	n:	P:

(ii) What do you conclude about the relationship between the frequency of visits and the amount an individual is prepared to pay?

5.3 For several makes of car, data for engine size and fuel economy are shown below:

Make	*Engine size (l)*	*Fuel economy (km l^{-1})*	*Make*	*Engine size (l)*	*Fuel economy (km l^{-1})*	*Make*	*Engine size (l)*	*Fuel economy (km l^{-1})*
A	1.0	14.3	I	1.1	14.9	Q	1.6	12.0
B	1.0	16.4	J	1.4	12.7	R	1.6	12.4
C	1.0	16.7	K	1.4	12.8	S	1.6	12.5
D	1.0	17.2	L	1.4	13.5	T	1.6	12.8
E	1.0	17.6	M	1.4	13.9	U	1.6	13.9
F	1.0	17.6	N	1.4	14.5	V	1.6	14.7
G	1.0	18.2	O	1.4	15.2	W	1.8	11.5
H	1.1	14.9	P	1.4	15.6	X	1.8	12.8

(i) To test whether fuel economy is dependent on the size of the engine, perform a regression analysis and record the following:

R^2:	Regression equation:	
F:	df:	P:

(ii) Plot these data with the regression line.

(iii) Plot the residuals against either the independent variable or the fitted values of y.

(iv) Does the distribution of the residuals meet with the assumptions of regression?

(v) What do you conclude about the relationship between engine size and fuel economy?

6 ▶ Analysing frequency data

This chapter covers tests for analysing frequency distributions of data measured on a nominal scale:

- Tests for association (or independence) between two variables measured on nominal scales
- Tests for goodness of fit between a measured and a theoretical distribution
- Consideration of alternative tests for association, goodness of fit and cases where there are more than two variables

So far we have considered the analysis of data that are measured at least on an ordinal scale (e.g. a ranking of tree leaf condition) or an interval/ratio scale (e.g. strontium-90 activity in becquerels per litre). We have only used nominal variables (i.e. data which are placed in categories) as a way of separating samples (e.g. males from females, urban from suburban, clean from polluted). We now need some way of analysing nominal scale data in their own right. Suppose that when we surveyed trees on polluted and clean sites (Chapter 4) we were unable to give the tree a rank score of condition and were merely able to say whether the tree was in poor condition or not. For each tree we would then have obtained a nominal measurement (good or poor) rather than a score from 1 to 6 (where good condition was 6). This would be the second piece of nominal data, because we also record site type: clean or polluted. How, then, would we analyse these data? We could compare the number of trees in good condition and the number of trees in poor condition on the two types of site using frequency analysis. There are two main types of frequency analysis: **tests for association** and tests for **goodness of fit**. The choice of test depends on whether we are examining the association of one measured frequency distribution with another (e.g. association of tree condition with site type), or whether we are testing one measured frequency distribution against a known theoretical distribution. The term 'theoretical distribution' sounds quite grand, but could simply be the expectation of equal frequencies of events (e.g. sex ratio: equal numbers of males and females).

Associations between frequency distributions

In a survey of public access agreements between local authorities and landowners, two variables were recorded: whether or not each authority had access agreements (in addition to public access via the public right-of-way network); and what type of local

Table 6.1 *Data sheet for survey into public access agreements made between local authorities and landowners*

Row (data record number)	Access agreements present (P) or absent (A)	Type of authority: district (D), metropolitan district (M), or London borough (L)
1	A	M
2	P	D
3	A	M
4	A	D
5	P	L
etc.	etc.	etc.
↓	↓	↓
287	P	M

authority it was (i.e. whether it was a district, a metropolitan district or a London borough). An extract from the data sheet is shown in Table 6.1, where each row is an authority, and 287 authorities were surveyed all together (i.e. there were 287 rows).

In this example, there are two **frequency distributions** (see Figure 6.1): the distribution of the presence or absence of access agreements and the distribution of authority types. Suppose we wish to know whether there is any association between these two distributions, i.e. whether the different authority types differ in their tendency to have additional access agreements with landowners. Another way of wording this is to say that we are examining whether the distributions are independent of each other (i.e. whether

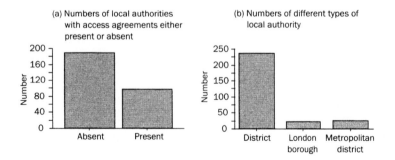

Figure 6.1 *Frequency distribution of types of local authority with access agreements either present or absent (n = 287)*

having an access agreement is independent of authority type): for this reason, this test is also known as a **test for independence**. The data records (from Table 6.1) can be summarised in the **contingency table** shown in Worked Example 6.1a. Here the frequency counts for all possible combinations (presence and absence of access agreements for each local authority type) are laid out in a table similar to that used for cross-tabulation in Chapter 2 (Table 2.7).

WORKED EXAMPLE 6.1A *Test for association between local authority types and presence or absence of access agreements: contingency table of observed values*

	Agreements present	Agreements absent	**Row totals**
Districts	68	169	**237**
Metropolitan districts	20	7	**27**
London boroughs	10	13	**23**
			Grand total
Column totals	**98**	**189**	**287**

The contingency table in Worked Example 6.1a is called a 3 × 2 contingency table (because there are three rows and two columns). Note that one distribution occupies the columns and the other the rows (it does not matter which way round these go). Since we are asking whether there is an association between the two distributions, the null hypothesis is that there is no association (i.e. that the distributions are independent of each other). We can then calculate what the expected values would be if authority type and the presence or absence of access agreements were distributed independently of each other. Or, worded another way, we calculate the frequencies expected if the distribution of access agreements was similar for each authority type. The first step is to work out the row and column totals (shown in Worked Example 6.1a), and then the expected frequencies are worked out by a simple formula which uses the row and column totals (see Box 6.1).

Note that, in analysing for associations between frequency distributions, most computer programs will accept the data either in a pair of columns, each containing the codes for a single distribution as in Table 6.1, or as a contingency table as in Worked Example 6.1a (for details of data entry for analysis by computer, see Tables B.4 and B.5, Appendix B). Note also that some programs require numerical codes rather than letters.

Worked Example 6.1b shows the calculation of each expected value. Note that, as a check on your arithmetic, the expected values should have the same row and column totals as the observed values, taking into account any rounding errors. We can see that

---

Box 6.1 *Formula for the expected values for testing associations between frequency distributions*

The expected values for each cell of a contingency table are calculated as:

$$\text{Expected value } (E) = \frac{\text{Row total} \times \text{Column total}}{\text{Grand total}}$$

This formula can be arrived at from basic probability theory (see Box 2.4). From the data in Worked Example 6.1a, there are 287 authorities in total. Of these, 237 are districts. If the null hypothesis is true, an authority randomly drawn from the 287 would have a 237/287 chance of being a district. Similarly with access agreements, 98 have access agreements, so there is a 98/287 chance of randomly drawing an authority with such agreements. If authority type and having access agreements are independent events, then the two probabilities are multiplied to find the probability of being a district *and* having access agreements:

$$\frac{237}{287} \times \frac{98}{287} = 0.282$$

This is a probability (and thus it lies between 0 and 1); to turn this into an expected number of districts with access agreements, we multiply the probability by the grand total (287):

$$\frac{237}{287} \times \frac{98}{287} \times 287 = 80.927$$

Note that this calculation simplifies to the general formula by cancelling 287 on the top and bottom of the equation:

$$\frac{237 \times 98}{287} = 80.927$$

---

there are differences between the expected and the observed values in Worked Example 6.1b, for example there are fewer districts with agreements than we would expect if authority type and presence of agreements were independent, and more metropolitan districts with agreements than expected. To identify whether these differences are significant, we need to calculate a test statistic to find whether this difference between the observed and expected frequencies is bigger than we would expect by chance. The statistic, X^2 (not to be confused with the lower case x^2 used in earlier chapters), is first calculated for each cell of the contingency table and then summed to obtain the total X^2 value (see Box 6.2). The distribution of this test statistic approximates to a statistical distribution known as chi-square (pronounced 'ky-square', where 'ky' rhymes with 'my', and given the symbol χ^2). In fact, this test is commonly known as a **chi-square test**. Many text books give the test statistic the symbol χ^2 instead of X^2, but strictly speaking this is not correct because X^2 only approximates to χ^2 (see Sokal and Rohlf, 1995).

For our access agreement data, the total X^2 value is 23.168 (see Worked Example 6.1c). To see if this is significant, we compare the total X^2 value to a table of chi-square values (see the shaded area of Table 6.2) at the appropriate degrees of

WORKED EXAMPLE 6.1B *Test for association between local authority types and presence or absence of access agreements: table of expected values (E). The observed values (O) are taken from Worked Example 6.1a. The formula for calculating E is (row total × column total)/grand total (see Box 6.1)*

	Agreements present			*Agreements absent*			*Row totals*
	O	*Calculating E*	*E*	*O*	*Calculating E*	*E*	
Districts	68	(237 × 98)/287 =	**80.927**	169	(237 × 189)/287 =	**156.073**	237
Metropolitan districts	20	(27 × 98)/287 =	**9.220**	7	(27 × 189)/287 =	**17.780**	27
London boroughs	10	(23 × 98)/287 =	**7.854**	13	(23 × 189)/287 =	**15.146**	23
Column totals	98			189		Grand total 287	

freedom, given, since we have three rows and two columns, by df = (3 − 1)(2 − 1) = 2 (see Box 6.2). At 2 degrees of freedom, our calculated value (23.168) is higher than the table values of chi-square for either $P = 0.05$ (5.99) or $P = 0.01$ (9.21), telling us that we have a highly significant association ($P < 0.01$) between authority type and presence or absence of access agreements.

Having demonstrated that there is a significant association between authority type and the presence of additional access agreements, we now return to the data to interpret these results. This can be done by looking at the individual cell X^2 components (see

Box 6.2 *Formula for the test statistic* X^2

The components of the test statistic are first calculated for each cell of the contingency table:

$$\text{Cell } X^2 = \frac{(O - E)^2}{E}$$

where: O is the observed (measured) value and E the expected (calculated) value.

The individual cell X^2 values are then summed to give the total X^2:

$$X^2 = \sum \frac{(O - E)^2}{E}$$

The total X^2 value is then compared with a table of chi-square values (χ^2: an extract from which is shown in Table 6.2) at df = $(r − 1)(c − 1)$, where r is the number of rows and c the number of columns: see Box 6.3 for an explanation of the logic behind this calculation of the degrees of freedom.

WORKED EXAMPLE 6.1C *Test for association between local authority types and presence or absence of access agreements: calculating the test statistic (X^2). Observed and expected values are taken from Worked Examples 6.1a and 6.1b, respectively. The formula for the cell X^2 values is $(O - E)^2/E$ (see Box 6.2)*

	Agreements present		Agreements absent	
	Calculating cell X^2 values	**Cell X^2**	*Calculating cell X^2 values*	**Cell X^2**
Districts	$(68 - 80.926)^2/80.926 =$	**2.0649**	$(169 - 156.073)^2/156.073 =$	**1.0707**
Metropolitan districts	$(20 - 9.220)^2/9.220 =$	**12.6058**	$(7-17.780)^2/17.780 =$	**6.5363**
London boroughs	$(10 - 7.854)^2/7.854 =$	**0.5866**	$(13 - 15.146)^2/15.146 =$	**0.3042**

Total X^2 = 2.0649 + 12.6058 + 0.5866 + 1.0707 + 6.5363 + 0.3042 = 23.168

This value is then compared with values in a chi-square table (see the shaded area of Table 6.2) at degrees of freedom calculated as:

df = $(r - 1)(c - 1) = (3 - 1)(2 - 1) = 2$

where r is the number of rows and c the number of columns.

Worked Example 6.1c). The biggest contributor to the total X^2 value stems from metropolitan districts (i.e. those covering towns and cities) having more than the expected number of additional access agreements. Interpretation of significant associations may depend on some knowledge of the material under study. In this example, towns and cities may have less access via the existing right-of-way network than do more rural areas, while having more people, thus creating a demand for additional access.

Contingency table analysis can be performed on tables of any size from 2 × 2 upwards (although 2 × 2 contingency tables are considered to be something of a special case: see below). However, as the table grows larger, the interpretation of any significant associations becomes more difficult.

Table 6.2 *Selected values of chi-square (χ^2). Reject the null hypothesis if the calculated value of X^2 is greater than the table χ^2 value. Shading indicates the critical values for the example referred to in the text. A more comprehensive table of χ^2 values is given in Table D.8 (Appendix D)*

df	χ^2	
	P = 0.05	*P = 0.01*
1	**3.841**	6.635
2	5.991	9.210
3	**7.815**	11.345
4	**9.488**	13.277
5	**11.070**	15.086

Box 6.3 *Logic underlying the calculation of degrees of freedom for a test of association*

When calculating the degrees of freedom for a test of association between frequency distributions, we do not use the number of observations, but rather the number of cells. In a 2 × 3 contingency table, there are 2 × 3 = 6 cells with a frequency in each. However, as for other statistical tests, the assumption of contingency table analysis is that each sample is independent. We used row and column totals to calculate our

2 × 3 contingency table			
	Present	Absent	Total
District	68	?	237
Metropolitan	20	?	27
London	?	?	23
Total	98	189	

expected values. For this size of table, we can see that with knowledge of the row and column totals, only two of the cells can be randomly drawn; the remaining cells are not independent, but can be calculated from the information that we already have. The degrees of freedom are therefore 2. This logic is extended for any size of table to give the formula:

$$df = (r - 1)(c - 1)$$

where r is the number of rows and c is the number of columns.

The special case of 2 × 2 contingency tables

When the degrees of freedom are 1 (e.g. in a 2 × 2 contingency table), the calculated X^2 value is not accurate (it is higher than it should be) and there is an increased chance of rejecting the null hypothesis when in fact it is true (i.e. concluding that there is an association when there is not: a type I error – see Chapter 3). An adjustment, called Yates' correction for continuity, can be made to the calculated value of X^2 for 2 × 2 contingency tables. This is not a perfect solution, since it tends to overcompensate and reduce the chances of obtaining a significant result to a greater extent than we would wish (i.e. giving an increased chance of a type II error). One way round this is to calculate the X^2 value both with and without the correction and compare the significance values obtained. If both agree, then you can be confident about the result. However, if the corrected value is not significant, but the uncorrected analysis gives a significant result, then the results are ambiguous and the true probability lies around 0.05. Under these circumstances, you should be cautious about your interpretation, and consider repeating the investigation using a larger sample size.

In Chapter 4, the condition of trees on two types of site (polluted and clean) was recorded on a scale of 1 to 6 (where 1 was poor condition). Suppose the researcher had instead classified the trees as 'poor' (for those with a score of 1, 2 or 3) or 'good' (for those with a score of 4, 5 or 6). These data can be entered into a 2 × 2 contingency table (see Worked Example 6.2). To apply the correction factor, simply subtract 0.5 from the absolute value of the $O - E$ component of the original formula before squaring it (see Box 6.4).

Box 6.4 *Formula for Yates' correction for continuity*

The formula for situations with 1 degree of freedom, using Yates' correction, is:

$$X^2 = \sum \frac{(|O - E| - 0.5)^2}{E}$$

where O are the observed values and E are the expected values. $|O - E|$ instructs you to take the absolute value of $O - E$. Remember that the absolute value ignores any negative sign and so is always positive.

WORKED EXAMPLE 6.2 *Test for association between tree condition and type of site*

First we complete a 2 × 2 contingency table by filling in the observed values. Expected values (in brackets) are calculated as (row total × column total) / grand total (see Box 6.2):

	Poor	*Good*	*Row totals*
Polluted	5 *(3)*	5 *(7)*	**10**
Clean	1 *(3)*	9 *(7)*	**10**
Column totals	**6**	**14**	**Grand total: 20**

Now we can calculate the test statistic, incorporating Yates' correction for continuity, as $\sum[(|O - E| - 0.5)^2/E]$ (see Box 6.4):

	Poor	*Good*
Polluted	$(|5 - 3| - 0.5)^2 / 3 = 0.75$	$(|5 - 7| - 0.5)^2 / 7 = 0.321$
Clean	$(|1 - 3| - 0.5)^2 / 3 = 0.75$	$(|9 - 7| - 0.5)^2 / 7 = 0.321$

Total $X^2 = 0.75 + 0.75 + 0.321 + 0.321 = 2.143$

When we look up the value calculated in Worked Example 6.2 in the chi-square table (Table D.8, Appendix D) at df = 1, we find that our value of 2.14 is less than the critical value (3.84), and there is therefore no significant association between the condition of trees and whether or not the site is polluted. If we had not applied the correction factor, our calculated value would have been 3.81 which, although still not significant, is close to the critical value. Since both the corrected and uncorrected values give a non-significant result, we can be confident that there is:

no significant association between the condition of the trees and the polluted nature of the site (corrected $X^2 = 2.14$, df = 1, $P > 0.05$).

It may seem surprising that these data are not significant using a test for associations between frequency distributions when there was a highly significant effect using a Mann–Whitney U test on the condition scores in Chapter 4 (see Worked Example 4.2). This example serves to illustrate a general point about downgrading data. By converting our scores to nominal categories (i.e. from an original range of scores from 1 to 6 to either 'good' or 'poor'), we have discarded much information from the original data, and thus lost sensitivity in detecting possible differences between the two types of site. In this example, had the researcher felt confident enough to give tree condition a rank score, valuable information would have been saved. In addition, the sample sizes here are quite low. Tests of association are more powerful when the sample size is large and more reliable when the mean expected frequency is large (see the section on the assumptions of chi-square frequency analysis later in this chapter).

Goodness of fit against theoretical distributions

When we wish to compare observed frequencies of one distribution with an external hypothesis, the frequency analysis is known as a goodness of fit test. Imagine we were interested in ascertaining whether the ethnic mix of visitors to a nature reserve was representative of that of the local community. In a survey of 500 visitors, 349 were classified as coming from ethnic group A, 117 as coming from ethnic group B and 34 as coming from ethnic group C (see the first two columns in Worked Example 6.3). These observed frequencies can be compared with expected frequencies based on known percentages of the different ethnic groups from census data: 64% (or 0.64 as a proportion) of the local population are of ethnic group A, 30% (0.3) are B and 6% (0.06) are C. Since this census information has not come from our study, it is the external hypothesis which is used to calculate the expected frequency distribution. The null hypothesis would be that there is no significant difference between the ethnic make-up of the visitors to the reserve and the make-up of the local community. The first step is to use the proportions from the census to generate the expected frequency of the ethnic groups in a sample of 500 (the fourth column in Worked Example 6.3). The expected values are calculated by multiplying the proportion expected from the census data by 500. Note that the sum of the expected values should also be 500.

The test statistic is again X^2 and is calculated using the same formula as before (Box 6.2). The calculation is shown in stages in the final three columns of Worked Example 6.3. For goodness of fit tests, the degrees of freedom are the number of categories minus 1. In this case there are 3 categories, so the degrees of freedom are 2. We look up our value of X^2 in a chi-square table at the appropriate degrees of freedom (see

Table D.8, Appendix D). Since our value (10.42) is higher than the table value at both $P = 0.05$ (5.99) and $P = 0.01$ (9.21), we can conclude that:

The frequency of the different ethnic groups at the reserve is not representative of the local population ($X^2 = 10.42$, df = 2, $P < 0.01$).

WORKED EXAMPLE 6.3 *Goodness of fit test between ethnicity of visitors to a nature reserve and the ethnic mix of the local community. The formula for calculating the expected values is shown in Box 6.1*

Ethnic group	Observed values (O)	Calculating the expected values (E)		E	O − E	(O − E)²	Cell X² values (O−E)² / E
A	349	500 × 0.64	=	320	29	841	2.628
B	117	500 × 0.30	=	150	−33	1089	7.260
C	34	500 × 0.06	=	30	4	16	0.533
Total	500			500			$X^2 = 10.42$

Looking at the final column of Worked Example 6.3, we can see that ethnic group B makes the greatest contribution to the overall X^2 value. When we compare the observed and expected values for this group, this suggests that people belonging to ethnic group B are underrepresented at the reserve.

Note that, in analysing for goodness of fit between an expected and observed frequency distribution, many computer programs will accept the data in a pair of columns, one for the expected and one for the observed distribution (for details of data entry for analysis by computer, see Tables B.4 and B.5, Appendix B).

This test can be applied irrespective of the number of categories, e.g. the frequency of ethnic groups A, B, C, D, E could be tested against an expected distribution (with four degrees of freedom: number of categories minus 1). However, the test may be unreliable if the expected frequencies fall below 2 (see the assumptions of chi-square frequency analysis listed later in this chapter). Goodness of fit tests with only two categories (i.e. the degrees of freedom are 1) are sometimes considered to be something of a special case (see below).

The special case of two-category goodness of fit tests

If we have two categories within which to place our data records, the degrees of freedom are 1, and there is a falsely enhanced chance of obtaining a significant result (type I error). Here, we can use a correction for continuity (as we did for 2 × 2 contingency tables). The formula for calculating X^2 is exactly the same as that for 2×2 contingency tables (see Box 6.4). Again, the correction overcompensates (gives an increased chance of a type II error) and so the same guidelines apply: the results can be taken with confidence unless the corrected analysis turns an initially significant result into a non-significant one. Under these circumstances it is worth repeating the study with a larger sample size.

A useful application of this two-category test is to test for equality in a sex ratio, with the null hypothesis (external hypothesis) being that there are 50% males and 50% females. Where the expected values are equal for each category, as in this case, the goodness of fit test is sometimes known as a test for **homogeneity** (i.e. a test for evenness). In a survey of invertebrate diversity using pitfall traps, an ecologist catches 91 spiders, only 33 of which are female. It seems that the trapping technique may be

biased towards catching males, but is it significant? Of the 91 spiders, if the sex ratio was $1 : 1$, we would have expected half (45.5) to be female. The calculation of X^2 (by Yates' correction formula given in Box 6.4) is given in Worked Example 6.4. Consulting the chi-square table at df = 1 (see Table D.8, Appendix D), we find that our calculated value of 6.33 lies between the table values for $P = 0.01$ $\chi^2 = 6.63$) and $P = 0.05$ ($\chi^2 = 3.84$), therefore we reject the null

hypothesis of sex ratio unity and conclude that the sex ratio is significantly biased towards males ($P < 0.05$). Note that if we had not used the correction factor, X^2 would have been 6.868 which is greater than the table value at $P = 0.01$ (6.63). Thus, the use of the correction alters the confidence we have in our result, but does not effect the main conclusion that:

the sex ratio of spiders caught in pitfall traps is significantly biased towards males (corrected $X^2 = 6.33$, df = 1, $P < 0.05$).

WORKED EXAMPLE 6.4 *Test for homogeneity of sex ratio in spiders caught using pitfall trapping*

	Observed O	Expected E	$O - E$	$(|O - E| - 0.5)^2$	Cell X^2 values $(|O - E| - 0.5)^2 / E$
Males	58	45.5	12.5	144	3.165
Females	33	45.5	12.5	144	3.165
					$X^2 = \mathbf{6.33}$

Goodness of fit against model distributions

Expected frequencies can be derived from a model distribution and compared to observed frequencies using a goodness of fit test. There are many possible distributions to which data might conform. In Chapter 2 we found that measurement data often conform to the normal distribution, and that an assumption of parametric tests (including the t test) is that data are normally distributed. A few statistics packages use goodness of fit tests to test the data against this assumption. However, it has been suggested that some alternative methods are better – see, for example, Zar (1999) for details.

Other theoretical distributions correspond to the way in which data are dispersed in time or space. For example, counts of plants, animals, people, buses, etc., may show them to be evenly (regularly) spaced, randomly spaced or clumped. If counts of relatively rare items are distributed randomly, then they may follow a distribution called the Poisson distribution (where the variance equals the mean). Regularly spaced items may conform to the binomial distribution (where the variance is small in comparison to the mean) and clumped items to the negative binomial distribution (where the variance is greater than the mean). Perfect distributions of these may be calculated and compared with the observed data using goodness of fit tests. See texts such as Zar (1999) or Sokal and Rohlf (1995) for details.

Assumptions of chi-square frequency analysis

In tests of association or goodness of fit the following assumptions should be satisfied:

1 The observed values should be counts (not proportions or percentages – these should be calculated back to their original frequencies before analysis).
2 The grand total must represent the number of independent individuals in the study.
3 Additionally, calculations of X^2 (as opposed to some other tests of association or goodness of fit – see the next section) are unreliable when the expected frequencies are low. In practice, when testing at a critical value of $P = 0.05$, the mean of the expected values (the grand total divided by the total number of cells in the table) should be at least 2 for goodness of fit tests (although if there are only two categories, each expected value should be at least 5) and at least 6 for tests of association (see Zar, 1999). If expected values are lower than this, then categories can be combined where appropriate, to increase the numbers of observations (and hence the expected values) in a particular category. If the critical value needs to be at P less than 0.05 (e.g. $P = 0.01$), then higher expected values are required (see Zar, 1999).

Fisher's exact test and G tests

There are alternatives to chi-square tests for frequency analysis, the best known of which are Fisher's exact test and the G test. Fisher's exact test is used only for 2×2 contingency tables where the research design has effectively fixed the values of both the row and column totals. It is not often appropriate for environmental and geographical data.

G tests can be used as an alternative to chi-square tests for most purposes, indeed some authors recommend G tests in preference to chi-square tests, although this has yet to become established in either textbooks or journal articles. G tests may perform slightly better than chi-square tests, especially if some of the expected frequencies are low (see point 3 under the assumptions of chi-square frequency analysis above). See texts such as Sokal and Rohlf (1995) for further details.

Testing for associations between more than two distributions

It may be that the ecologist who collected the data for tree condition on polluted and clean sites in Worked Example 6.2 decided to repeat the survey on additional polluted and clean sites. It could then be tempting to combine the new data with the existing data into one contingency table. However, this would assume that the distributions of tree condition and site type were drawn from the same underlying frequency distributions on both occasions. It is only permissible to combine the data for the two (or more) occasions if the two (or more) tests are first run individually and then tested for consistency (see Zar, 1999, for details). If they are shown to be consistent, the data can then be combined and analysed using either a chi-square or a G test.

If a third variable in addition to site type and tree condition had been measured (e.g. different species of tree), then the contingency table could be viewed as being three-dimensional. This type of research design can be analysed using a G test (see Sokal and Rohlf, 1995, for details).

Summary

By now you should be able to:

- recognise when frequency data are appropriate for contingency table analysis (for association between two variables measured on nominal scales), calculate the expected values and work out the degrees of freedom;
- calculate the test statistic (X^2) for association between frequency distributions from the observed and expected values and assess its significance;
- recognise when data are appropriate for a goodness of fit test (i.e. to test frequencies against an expected external distribution), calculate expected frequencies from probabilities and work out the degrees of freedom;
- calculate X^2 for goodness of fit from the observed and expected values and assess its significance;
- interpret the direction of any significant X^2 (whether for association or goodness of fit) by looking at the individual X^2 components.
- be aware that other techniques are appropriate for the association between more than two frequency distributions.

Questions

6.1 As part of a study into the effects of woodland management on bird distribution, 36 woods were surveyed for the presence (P) or absence (A) of long-tailed tits and classified as one of three woodland types: pre-thicket (PT), thicket (T) or high forest (HF). The following data were obtained:

Wood number	Wood type	Presence or absence of long-tailed tits	Wood number	Wood type	Presence or absence of long-tailed tits
1	PT	P	19	HF	A
2	HF	P	20	T	P
3	PT	A	21	PT	P
4	PT	P	22	PT	P
5	HF	A	23	HF	A
6	T	P	24	T	P
7	HF	A	25	PT	A
8	T	P	26	PT	P
9	HF	P	27	HF	P
10	T	P	28	T	P
11	PT	A	29	PT	A
12	T	A	30	PT	P
13	T	P	31	T	P
14	HF	P	32	HF	A
15	PT	A	33	T	P
16	T	P	34	HF	A
17	HF	A	35	PT	P
18	HF	P	36	HF	A

(i) Arrange the data in a 3 × 2 contingency table (including row, column and grand totals).
(ii) Calculate the expected frequencies, place them in a table and check that the row, column and grand totals agree with those for the table of observed values.
(iii) Calculate the individual X^2 components and place them in a table.
(iv) Calculate the total X^2, df and P values.
(v) What do you conclude regarding the distribution of long-tailed tits in the different types of woodland?

6.2 Trees were surveyed for the presence of lichen. Each lichen found was recorded as either occurring on the north, east, south or west quadrant of the tree: 36 lichens were found on north-facing quadrants, 50 on east-facing quadrants, 29 on south-facing quadrants and 45 on west-facing quadrants.

(i) For lichens occurring on the north, east, south and west sides of trees, use the external hypothesis that lichen orientation is random, and display in a table the observed and expected number of lichens and the component X^2 values in a goodness of fit test.
(ii) Calculate the total X^2, df and P values to complete the goodness of fit test.
(iii) What can you conclude about the distribution of lichens on trees?

7 Differences between more than two samples

To investigate potential differences between more than two samples we need an extension of the tests described in Chapter 4. The tests are classified not only on their requirement for the data to be normally distributed, but also according to whether we are looking at more than one independent variable at a time. This chapter covers:

- **One-way analysis of variance for normally distributed data**
- **Kruskal–Wallis one-way analysis of variance using ranks for data measured at least on an ordinal scale**
- **Two-way analysis of variance for normally distributed data**
- **Two-way analysis of variance using ranks for data measured at least on an ordinal scale**
- **Consideration of alternative models of analysis of variance for more complicated experimental and survey designs**

In Chapter 4 we compared the means (or medians) of two samples using a *t* test (for normally distributed data) or a Mann–Whitney *U* test (for ordinal or non-normal data). However, many experiments or surveys involve the comparison of three or more samples. For example, in a survey of household energy bills, we might place houses in three categories depending on the type of insulation they have – loft insulation, double-glazing or minimal insulation. To investigate the effect of insulation type on household energy bill, it might seem logical to perform three *t* tests (comparing 'loft' with 'double-glazing', 'loft' with 'minimal' and 'double-glazing' with 'minimal'). This would certainly be possible. However, if we had more samples, the computation would become very long-winded and time-consuming (for seven samples, for example, we would need 21 separate tests). There is a more serious

objection to conducting multiple *t* tests: that is, when we compute several tests we increase our chances of obtaining a significant result by chance alone. Remember that the critical probability value is 0.05. This is the value at which the null hypothesis is rejected because there is only a 5% probability that the result occurred by chance. We can turn this around and say that on 5% of occasions the null hypothesis is rejected when in fact it is true (this is a type I error: see Chapter 3). It now becomes clear that if we do 20 tests, at least one may appear to be significant even if there were no real differences between the means.

To get around this problem, we need a test that will simultaneously compare three or more means (or medians) with an overall 5% probability of committing a type I error. For normally distributed data, we use a parametric **analysis of variance (ANOVA)**. There are several types of ANOVA depending on the experimental or survey design: the simplest type is used to compare two or more means and is called a **one-way analysis of variance**. In one-way ANOVA there is one independent variable (in our household energy example, insulation type) and one dependent variable (energy bill). An assumption of one-way ANOVA is that all the data are randomly collected and each data point is independent of all the others. This assumption would not be true in our energy bill example if half the houses had been sampled from London and half from Manchester, because there are likely to be differences between the two areas (e.g. in climate) that would affect energy bills. In this case, the survey would have two independent variables: insulation type and city. The appropriate way of analysing experiments or surveys with two independent variables is by **two-way analysis of variance**, which will be considered later in this chapter.

If the data cannot be used in a parametric analysis of variance, either because they are measured on an ordinal scale, or are not normally distributed and cannot be transformed, then a nonparametric version of one-way ANOVA, **Kruskal–Wallis one-way analysis of variance using ranks**, should be applied. When there is more than one independent variable, then an equivalent nonparametric two-way procedure is used.

This chapter first covers one-way analyses of variance (parametric and then nonparametric tests) and then two-way analyses (again both parametric and nonparametric tests). Extensions of the matched-pair tests of Chapter 4 follow. Finally, analyses that are appropriate for more complex experimental and survey designs are briefly considered.

Parametric one-way analysis of variance

As in many of the tests covered in this book, ANOVA compares the two major sources of variation in the data: variation that is explained by the design of the experiment or survey (explained variation) and that which is due to the inherent variation within the samples (unexplained variation). In the energy bill example, the aim is to investigate whether variation in energy bill can be explained by type of insulation. Of course, variation in energy bill could be caused by many factors, for example type of heating system and how many hours the house is heated. Although in the survey we would standardise as many factors as possible (for example by selecting houses of approximately similar size occupied by similar numbers of people), there would still be inherent variation in the energy bills. However, if insulation type is an important factor in explaining variation in energy bills, we would expect the explained variation (due to insulation type) to be greater than the unexplained variation (inherent variation). One-way ANOVA effectively calculates the test statistic (F) as the ratio of explained to unexplained variation (the F statistic is sometimes called the variance ratio).

It is an assumption of ANOVA that the variances of the samples are not different, therefore the first step of the analysis is to check this. If the variances are significantly different, then transforming the data by one of the methods in Chapter 3 may equalise the variances (this is because the mean of some non-normal distributions is dependent on the variance, and transformation can simultaneously remove this dependence and normalise the data: see Chapter 3). If, following transformation, the variances still differ, then we would need a nonparametric test (the Kruskal–Wallis one-way analysis of variance using ranks, which is covered in the next section).

Testing for equality of variances

The test for equality of variances (also known as homoscedasticity) is simple. First, the variances are obtained for each sample. The largest variance is then compared with the smallest (because if the largest and smallest are not significantly different from each other then the others cannot be). The largest variance is divided by the smallest to obtain a value called F_{max}, and this is compared to a table of critical values of F_{max} (an extract from which is given in Table 7.1) to find the probability of there being no significant difference between the variances. This procedure is demonstrated in Worked Example 7.1a on energy bill data from households with three different insulation types. If the calculated value equals or is lower than the table value at $P = 0.05$ (i.e. $P \geq 0.05$), we can accept the null hypothesis that there is no significant difference between the variances and proceed with the ANOVA. If our value is greater than the table value, then the variances are significantly different ($P < 0.05$) and we should either transform the data (if appropriate) and recheck, or use a nonparametric test. The table is consulted at two points: k, the number of samples we are comparing (in this case 3) and the degrees of freedom ($n - 1$) of the samples (in this case there are eight houses in each sample, so the degrees of freedom are 7). If we had an unequal number of data points in the two samples being compared by the F_{max} test then we would use the degrees of freedom of the sample with fewer data points. In our example, where $k = 3$ and df $= 7$, the critical value of F_{max} at $P = 0.05$ is 6.94 (see the shaded area in Table 7.1). From Worked Example 7.1a, the calculated F_{max} (1.109) is lower than this, so we can conclude that the variances are not significantly different ($P > 0.05$) and it is safe to proceed with the ANOVA.

Note that the data in Worked Example 7.1a are listed in three separate columns (one for each sample). If you use a computer program to calculate your ANOVA tests, this format of data will often not be appropriate. See Appendix B (Table B.6) for details of data entry for statistical analysis using computers.

WORKED EXAMPLE 7.1A *ANOVA for energy bills (£ per annum) in houses with different types of insulation: calculating the F_{max} test*

Loft insulation, x_L ($n = 8$)	Double glazing, x_D ($n = 8$)	Minimal, x_M ($n = 8$)
149	134	171
198	181	184
179	198	191
158	187	213
134	151	198
155	174	186
181	178	234
161	169	199

Σx_L =	1 315	Σx_D =	1 372	Σx_M =	1 576
Σx_L^2 =	219 073	Σx_D^2 =	238 212	Σx_M^2 =	313 104
$\bar{x}_L$ =	164.4	$\bar{x}_D$ =	171.5	$\bar{x}_M$ =	197
s_L^2 =	417.1	s_D^2 =	416.3	s_M^2 =	376

$$F_{max} = \frac{\text{Largest variance}}{\text{Smallest variance}}$$

$$F_{max} = \frac{s_L^2}{s_M^2} = \frac{417.13}{376.00} = 1.109$$

At df = 7 and k = 3, from the shaded area of Table 7.1, we can see that *P* > 0.05

Table 7.1 *Selected values of F$_{max}$ for testing equality of variances. Reject the null hypothesis if the calculated value of F$_{max}$ is greater than the table value. Values are given for P = 0.05. Shaded lines indicate the critical values for the example referred to in the text. A more comprehensive table of F$_{max}$ values is given in Table D.9 (Appendix D)*

		Number of samples being compared (k)			
		2	3	4	5
Degrees of freedom	6	5.82	8.38	10.4	12.1
of the smallest	7	4.99	6.94	8.44	9.70
sample (n – 1)	8	4.43	6.00	7.18	8.12

Calculating the *F* statistic for ANOVA

Analysis of variance involves first finding the total variation in the data (measured as the **sums of squares**: first encountered in step 3 in Box 2.3) and dividing this variation into explained variation (i.e. that which is accounted for by the category or treatment – in this case insulation type – which is also called between variation, or $SS_{between}$) and unexplained variation (the inherent variation within the samples, also called within variation, or SS_{within}). Thus:

$$SS_{total} = SS_{between} + SS_{within}$$

The formulae for calculating the sums of squares are shown in Box 7.1, and the calculations for the energy example are shown in Worked Example 7.1b.

Box 7.1 *Formulae for the sums of squares for ANOVA*

The sum of squares of all the data points is:

$$SS_{total} = \sum x_T^2 - \frac{\left(\sum x_T\right)^2}{n_T}$$

where:

$\sum x_T$ is the sum of all values in all samples;

n_T is the total number of values in all samples.

The within variation is the sum of the variation in the individual samples. When comparing the means of k samples, the within sum of squares is:

$$SS_{within} = SS_1 + SS_2 + SS_3 + \ldots + SS_k$$

where:

$$SS_1 = \sum x_1^2 - \frac{\left(\sum x_1\right)^2}{n_1}$$

The between sample sum of squares is calculated by:

$$SS_{between} = \frac{\left(\sum x_1\right)^2}{n_1} + \frac{\left(\sum x_2\right)^2}{n_2} + \frac{\left(\sum x_3\right)^2}{n_3} + \cdots + \frac{\left(\sum x_k\right)^2}{n_k} - \frac{\left(\sum x_T\right)^2}{n_T}$$

Here $\sum x_1$ is the sum of all values in sample 1, $\sum x_2$ the sum of all values in sample 2, and so on up to sample k; n_1 is the number of values in sample 1, n_2 the number of values in sample 2, and so on up to sample k.

As a check, the total sum of squares should be equal to the between plus the within sums of squares:

$$SS_{total} = SS_{between} + SS_{within}$$

Note that these formulae are the same irrespective of whether sample sizes are equal or unequal.

WORKED EXAMPLE 7.1B *ANOVA for energy bills (£ per annum) in houses with different types of insulation: calculating the sums of squares using the data from Worked Example 7.1a*

Loft ($n = 8$)		Double-glazing ($n = 8$)		Minimal ($n = 8$)		Total ($n = 24$)	
$\sum x_L$	$=$ 1315	$\sum x_D$	$=$ 1372	$\sum x_M$	$=$ 1576	$\sum x_T$	$=$ 4263
$\sum x_L^2$	$=$ 219 073	$\sum x_D^2$	$=$ 238 212	$\sum x_M^2$	$=$ 313 104	$\sum x_T^2$	$=$ 770 389
$\bar{x}_L$	$=$ 164.4	$\bar{x}_D$	$=$ 171.5	$\bar{x}_M$	$=$ 197	$\bar{x}_T$	$=$ 177.625
s_L^2	$=$ 417.1	s_D^2	$=$ 416.3	s_M^2	$=$ 376	s_T^2	$=$ 572.766

First we compute the total sum of squares:

$$SS_{total} = \sum x_T^2 - \frac{\left(\sum x_T\right)^2}{n_T} = 770\,389 - \frac{(4263)^2}{24} = 13\,173.625$$

and then the sum of squares for each individual sample:

$$SS_{loft} = \sum x_L^2 - \frac{\left(\sum x_L\right)^2}{n_L} = 219\,073 - \frac{1315^2}{8} = 2919.875$$

$$SS_{double\text{-}glazing} = \sum x_D^2 - \frac{\left(\sum x_D\right)^2}{n_D} = 238\,212 - \frac{1372^2}{8} = 2914$$

$$SS_{minimal} = \sum x_M^2 - \frac{\left(\sum x_M\right)^2}{n_M} = 313\,104 - \frac{1576^2}{8} = 2632$$

Now we calculate SS_{within} and $SS_{between}$:

$$SS_{within} = SS_{loft} + SS_{double\text{-}glazing} + SS_{minimal} = 2919.875 + 2914 + 2632 = 8465.875$$

$$SS_{between} = \frac{\left(\sum x_L\right)^2}{n_L} + \frac{\left(\sum x_D\right)^2}{n_D} + \frac{\left(\sum x_M\right)^2}{n_M} - \frac{\left(\sum x_T\right)^2}{n_T}$$

$$= \frac{1315^2}{8} + \frac{1372^2}{8} + \frac{1576^2}{8} - \frac{4262^2}{24} = 761\,923.125 - 757\,215.375 = 4707.75$$

As a check on your calculations, you can confirm that:

$$SS_{total} = SS_{within} + SS_{between}$$

$$13173.625 = 8465.875 + 4707.75$$

Table 7.2 *Producing an ANOVA table*

	Column				
	1	2	3	4	5
Source of variation	df [a]	Sums of squares (SS)	Mean squares (MS)	F statistic	P
Between	$k - 1$	$SS_{between}$	$\dfrac{SS_{between}}{df_{between}}$	$\dfrac{MS_{between}}{MS_{within}}$	Look up in tables using the between df and the within df
Within	$(n_T - 1) - (k - 1)$	SS_{within}	$\dfrac{SS_{within}}{df_{within}}$		
Total	$n_T - 1$	SS_{total}			

[a] k is the number of samples being compared; n_T is the number of data points in the whole analysis.

The sums of squares values are now used to calculate the test statistic (F). These calculations can be arranged in an ANOVA table as in Table 7.2 (this is also the usual output from statistics programs). The degrees of freedom are entered in column 1: the total degrees of freedom are the total number of data points in the whole analysis minus 1; the between degrees of freedom are the number of samples minus 1; and the within degrees of freedom are the total df minus the between df. The sums of squares that we calculated in Worked Example 7.1b are entered in column 2. Note that the between and within degrees of freedom add up to the total degrees of freedom, and the between and within sums of squares add up to the total sum of squares. In column 3, the **mean squares** (abbreviated to *MS*) are simply the sums of squares divided by their respective degrees of freedom to standardise the variation. In column 4, the test statistic, F, is calculated as $MS_{between}$ divided by MS_{within}. This is the standardised between variation (explained) divided by the standardised within variation (unexplained). The test statistic is then compared to table values of F (an extract of which is shown in Table 7.3) using two values of degrees of freedom (the between and the within) to give the probability of there being no significant difference between the means (column 5). Several alternative terms for the components of the ANOVA table are used in statistics textbooks and statistical software. The between variation is sometimes called treatment, factor or group variation, the within variation is sometimes called error (from the Latin *errare* – to wander) or residual (remaining variation). The ANOVA table for our energy example, including the calculations, is laid out in Worked Example 7.1c.

Once we have obtained the F value, we need to decide whether it is significant. Remember that the F value is the ratio of the explained to the unexplained variation. The null hypothesis is that the explained variation (between variation) is not significantly greater than the unexplained variation (within variation). The alternative hypothesis is that the explained variation is greater than the unexplained variation (i.e.

Table 7.3 *Selected values of the F distribution for ANOVA. Reject the null hypothesis if the calculated values of F are greater than the table values. The upper (bold) figures are for P = 0.05, while the lower figures are for P = 0.01. Shading indicates the critical values for the example referred to in the text. A more comprehensive table of F values is given in Table D.7 (Appendix D)*

		df for the between MS			
		1	2	3	4
	19	4.38	3.52	3.13	2.90
		8.18	5.93	5.01	4.50
	20	4.35	3.49	3.10	2.87
		8.10	5.85	4.94	4.43
df for the within MS	21	4.32	3.47	3.07	2.84
		8.02	5.78	4.87	4.37
	22	4.30	3.44	3.05	2.82
		7.95	5.72	4.82	4.31
	23	4.28	3.42	3.03	2.80
		7.88	5.66	4.76	4.26

WORKED EXAMPLE 7.1C *ANOVA for energy bills (£ per annum) in houses with different types of insulation: calculating the F statistic using the sums of squares from Worked Example 7.1b*

	df	Sums of squares (SS)	Mean squares (MS)	F statistic	P
Between	$3 - 1 = 2$	4 707.75	$\frac{4707.75}{2} = 2353.875$	$\frac{2353.875}{403.1369} = 5.839$	Look up
Within	$(24 - 1) - (3 - 1) = 21$	8 465.875	$\frac{8465.875}{21} = 403.1369$		in tables at df of 2
Total	$(24-1) = 23$	13 173.625			and 21

the F value is greater than 1, as is the case for our energy bill data – see Worked Example 7.1c, where the F value is 5.839). To see whether the explained variation is significantly larger than the unexplained variation, we compare our calculated F value to a table of F values, using the degrees of freedom for between (df = 2) and within (df = 21), as shown in the shaded area of Table 7.3. In cases where the explained variation is lower than the unexplained variation (i.e. the F value is less than 1), we can automatically accept that the null hypothesis is correct (i.e. there is no significant difference between the means) and $P > 0.05$. Looking at Table 7.3, for 2 and 21 degrees of freedom, our calculated F value (5.839) exceeds the table value for both $P = 0.05$ ($F = 3.47$) and $P = 0.01$ ($F = 5.78$). We therefore reject the null hypothesis and accept the alternative hypothesis that the variation between insulation categories is greater than the variation within insulation categories. The significant difference between the means could be expressed in the statement:

Insulation category (whether loft insulation, double-glazing or minimal insulation) has a highly significant effect on household energy bills ($F_{2,21} = 5.839$, $P < 0.01$).

Do not forget to quote both values of the degrees of freedom: notice the abbreviated way of expressing the degrees of freedom as a subscript to the F symbol with the between degrees of freedom first, followed by the within degrees of freedom. If you prefer, an abbreviated version of the ANOVA table (such as that shown in Table 7.4) can be placed in the results section of a report. Notice that we do not need to put in the sums of square values or the total degrees of freedom because the reader has enough information to calculate them.

Table 7.4 ANOVA for the effect of insulation type on energy bills

Source	df	Mean squares	F	P
Between	2	2353.875	5.839	< 0.01
Within	21	403.137		

We have demonstrated ANOVA on the comparison of three means. The test can be used to compare any number of means, including just two means: i.e. the situation where we used a t test in Chapter 4. If F is computed instead of t we obtain exactly the same probability, because the t and F tests are mathematically related (in fact, for comparisons of two samples $F = t^2$).

After obtaining a significant F value in an ANOVA on more than two means, we can conclude that there is a significant difference between the means, but we cannot say where these differences lie. With two samples we could simply examine the means (if there is a significant difference, the larger mean must be significantly larger than the smaller one). With three or more samples, we can view the means and get an impression: in our example, houses with minimal insulation have the most expensive bills (£197.00 – see Worked Example 7.1a) and those with loft insulation the least expensive (£164.40) with those with double-glazing being somewhere in the middle (£171.50), but we cannot say that all of the means differ significantly from each other. In fact, there are several possibilities:

Minimal > Double-glazing > Loft
Minimal > (Double-glazing, Loft)
(Minimal, Double-glazing) > Loft

where the samples in brackets are not significantly different from each other.

To tell where the differences lie we need a further test, known as a multiple comparison test. Where there are equal sample sizes (as in this example), the Tukey test is appropriate. If sample sizes differ then we need to use a Tukey–Kramer test. Note that some computer programs do not calculate Tukey or Tukey–Kramer tests.

Multiple comparison tests with equal sample sizes: the Tukey test

Multiple comparison tests are used only after an overall significant difference has been demonstrated. The test is done on all possible pairs of samples. To calculate how many tests need to be computed for k samples, the formula is:

$$\frac{k(k-1)}{2}$$

There are many methods for calculating multiple comparisons which all attempt to get around the problem of conducting several non-independent tests (for three samples there are $(3 \times 2) / 2 = 3$ tests, for four samples there are $(4 \times 3) / 2 = 6$ tests, etc.). This is a problem because the more tests we do, the more chance we have of falsely rejecting a null hypothesis and accepting a difference where one does not exist (type I error: see Chapter 3). Ideally we should compare the samples at an overall significance level of 5%. However by doing so, we increase the chances of falsely rejecting significant differences (type II error). The test we will use now is called the Tukey test (or HSD – honestly significant difference test), and appears to be an acceptable compromise between the two types of error. Box 7.2 shows the formula. The first step is to calculate the minimum significant difference (MSD) between samples. The MSD is then compared to the absolute differences between the pairs of means and those which are equal to or greater than the MSD are significantly different.

Box 7.2 Formula for the Tukey multiple comparison test for use after parametric ANOVA (where sample sizes are equal)

Here, the minimum significant difference is calculated as:

$$MSD = q \times SE$$

where the standard error is:

$$SE = \sqrt{\frac{MS_{within}}{n}}$$

and:

MS_{within} is the mean square for the within variation taken from the ANOVA table;

n is the number of values in each sample;

q is the value obtained from q distribution tables (see Table D.10, Appendix D) using the number of samples involved (k) and the degrees of freedom for the within sample category.

The calculation of the MSD involves a new value (q) which is multiplied by a standard error term. This standard error term is the square root of the within mean square from the ANOVA table (in our example, the value of 403.137 from Table 7.4) divided by the number of data points in each sample being compared (in our example there are eight houses in each sample). The appropriate value of q is taken from a table (an extract of which is shown in Table 7.5) against the number of samples being compared (in our example, 3) and the degrees of freedom for the within sample category in the ANOVA table (in this case 21 – see Table 7.4). Since the exact number of degrees of freedom for our example (21) are not covered by the q table (see the shaded area of Table 7.5), Worked Example 7.1d shows a method of estimating it, called **interpolation**. An alternative, conservative method (i.e. one that increases the chances of a type II error) is simply to take the q value for the next lowest degrees of freedom (i.e. 20).

WORKED EXAMPLE 7.1D *ANOVA for energy bills (£ per annum) in houses with different types of insulation: calculating the minimum significant difference between means*

To calculate the minimum significant difference (MSD) between our samples, we first calculate the *SE*:

$$SE = \sqrt{\frac{MS_{within}}{n}} = \sqrt{\frac{403.137}{8}} = 7.0987$$

The MSD is simply this value multiplied by q for degrees of freedom (= 21) and number of means (= 3) from Table 7.5. Since we cannot find the exact q value for degrees of freedom of 21, we interpolate between the values for 20 and 24 degrees of freedom. From Table 7.5 the value for df = 21 must lie between 3.58 (the value for df = 20) and 3.53 (the value for df = 24). For the 4 degrees of freedom between the table values, the value of q decreases by 3.58 – 3.53 = 0.05 units. If we assume that there are equally sized increments between the values on the table, then there is a drop of 0.05 / 4 = 0.0125 units for each degree of freedom between 20 and 24. To find the value at 21 degrees of freedom, one increment is removed from the value at df = 20 (i.e. 3.58 – 0.0125 = 3.5675). We can now calculate the MSD:

$$MSD = q \times SE = 3.5675 \times 7.0987 = 25.32$$

For the means to differ significantly from one another, they must differ by at least 25.32. We now compile a table giving the differences between the mean energy bill per insulation type (mean energy bills for each sample – in £ per annum – are given in brackets; differences between means that are greater than the MSD are highlighted with an asterisk):

	Loft (164.375)	Double-glazing (171.500)
Double-glazing (171.500)	7.125	–
Minimal (197.000)	32.625 *	25.500 *

Table 7.5 *Selected values of q for the Tukey test. Table values are for P = 0.05. Shading indicates the critical values for the example referred to in the text. A more comprehensive table of q values is given in Table D.10 (Appendix D)*

	Number of samples being compared		
	3	4	5
Degrees of	19	3.59	3.98	4.25
freedom for	20	3.58	3.96	4.23
within *MS*	24	3.53	3.90	4.17
	30	3.49	3.85	4.10

The goal of the test is to determine whether any of the absolute differences between the means are greater than the MSD. These comparisons for our energy and insulation data are shown in Worked Example 7.1d. Two of the comparisons exceed the MSD and are therefore significantly different from each other. In a results section of a report, this can be expressed as:

Insulation category (whether loft insulation, double glazing or minimal insulation) has a highly significant effect on household energy bills ($F_{2,21}$ = 5.839, $P < 0.01$). Multiple comparisons using the Tukey test revealed that the minimally insulated households have significantly higher bills than those with either loft insulation or double glazing ($P < 0.05$). However, there is no significant difference between the energy bills of houses with loft insulation and those with double glazing ($P > 0.05$).

Alternatively the following code could be used (this is a useful shorthand if summarising many results in one table):

Minimal > (Loft insulation, Double-glazing)

where items in brackets indicate no significant difference between them. Another method of displaying the Tukey test results is to add them to a graphical illustration of the means using symbols, as shown in Figure 7.1.

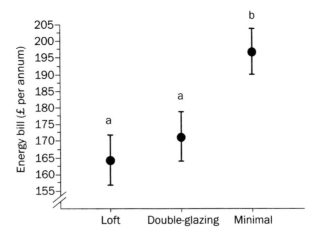

Figure 7.1 *Annual energy bills (mean ± SE) in households with different types of insulation (n = 8 in each sample). Means labelled with the same letter are not significantly different from each other following the Tukey test for multiple comparisons*

Note that it is possible to find a significant F value (meaning that at least one mean differs from the others) and yet find no significant multiple comparisons. This may seem odd, but it arises because multiple comparisons are not very sensitive tests.

Multiple comparisons with unequal sample sizes: the Tukey–Kramer test

For multiple comparisons of samples of unequal sizes, the Tukey–Kramer method should be used. The formula is very similar to the Tukey test (in fact if sample sizes are equal, the formula simplifies to the Tukey formula). It is more laborious to calculate by hand, because the minimum significant difference between each pair of means is calculated for each pair in turn using the formula in Box 7.3.

Box 7.3 Formula for the Tukey–Kramer multiple comparison test for use after parametric ANOVA (where sample sizes are not equal)

Here the minimum significant difference is calculated for each pair of means in turn:

$$\text{MSD}_{A,B} = q \times SE_{A,B}$$

where the standard error is:

$$SE_{A,B} = \sqrt{\frac{MS_{within}}{2}\left[\frac{1}{n_A} + \frac{1}{n_B}\right]}$$

and:

$\text{MSD}_{A,B}$ is the minimum significant difference between the means of samples A and B;

n_A and n_B are the number of items in samples A and B;

q is the value obtained from q distribution tables (see Table D.10, Appendix D) using the total number of samples involved (k) and the degrees of freedom for the within sample category.

To illustrate the Tukey–Kramer test, Worked Example 7.2 shows measurements of organochlorine residues in gull eggs from colonies situated in four lakes (A, B, C and D). Lakes A, B and D have eight replicate measurements, but lake C has only six. The data are analysed using ANOVA and, since the probability is less than 0.05, the analysis is completed by performing multiple comparisons using the Tukey–Kramer method for unequal sample sizes. From these analyses we find that C > (D, A) > B. We can report this as:

WORKED EXAMPLE 7.2 *ANOVA followed by the Tukey–Kramer test for minimum significant differences on organochlorine residues (mg kg⁻¹) in gull eggs*

Lake A (n = 8)	Lake B (n = 8)	Lake C (n = 6)	Lake D (n = 8)	Total (n = 30)
3.6	2.5	5.8	3.0	
3.0	2.4	6.2	3.3	
3.4	2.6	6.5	3.7	
2.9	2.6	6.1	3.6	
2.9	2.8	5.6	3.5	
3.1	2.5	6.3	3.6	
3.5	2.7		3.2	
3.3	2.6		3.4	

$$\Sigma x_A = 25.7 \quad \Sigma x_B = 20.7 \quad \Sigma x_C = 36.5 \quad \Sigma x_D = 27.3 \quad \Sigma x_T = 110.2$$

$$\Sigma x_A^2 = 83.09 \quad \Sigma x_B^2 = 53.67 \quad \Sigma x_C^2 = 222.59 \quad \Sigma x_D^2 = 93.55 \quad \Sigma x_T^2 = 452.9$$

$$\bar{x}_A = 3.2125 \quad \bar{x}_B = 2.5875 \quad \bar{x}_C = 6.083\,333 \quad \bar{x}_D = 3.4125 \quad \bar{x}_T = 3.673\,333$$

$$SS_A = 0.528\,75 \quad SS_B = 0.108\,75 \quad SS_C = 0.548\,333 \quad SS_D = 0.388\,75 \quad SS_T = 48.098\,67$$

$$s_A^2 = 0.075\,54 \quad s_B^2 = 0.015\,536 \quad s_C^2 = 0.109\,667 \quad s_D^2 = 0.055\,536 \quad s_T^2 = 1.658\,575$$

The variances were not significantly different (checked using an F_{max} test), therefore we can proceed with the ANOVA (using the formulae in Box 7.1 for sums of squares and Table 7.2 for the other components):

	df	SS	MS	F	P
Between	3	46.524 083 333	15.508 027 778	256.073	< 0.01
Within	26	1.574 583 333	0.060 560 897		
Total	29	48.098 666 667			

There is an overall significant difference between the lakes. To find where the differences lie, multiple comparisons are needed. There are $k(k-1)/2 = (4 \times 3)/2 = 6$ possible comparisons. First we calculate the standard error (SE) for each pair:

$$SE_{A,B} = \sqrt{\frac{MS_{within}}{2}\left[\frac{1}{n_A} + \frac{1}{n_B}\right]} = \sqrt{\frac{0.060\,560\,897}{2}\left[\frac{1}{n_A} + \frac{1}{n_B}\right]} = \sqrt{0.030\,280\,448\left[\frac{1}{n_A} + \frac{1}{n_B}\right]}$$

For each comparison the SE is as follows:

A (n = 8) vs B (n = 8)

$$\sqrt{0.030\,280\,448\left[\frac{1}{8} + \frac{1}{8}\right]} = 0.087\,006\,391$$

A (n = 8) vs C (n = 6)

$$\sqrt{0.030\,280\,448\left[\frac{1}{8} + \frac{1}{6}\right]} = 0.093\,977\,643$$

continued . . .

Worked Example 7.2 continued . . .

A ($n = 8$) vs D ($n = 8$)

$$\sqrt{0.030\,280\,448\left[\frac{1}{8}+\frac{1}{8}\right]} = 0.087\,006\,391$$

B ($n = 8$) vs C ($n = 6$)   B ($n = 8$) vs D ($n = 8$)

$$\sqrt{0.030\,280\,448\left[\frac{1}{8}+\frac{1}{6}\right]} = 0.093\,977\,643 \qquad \sqrt{0.030\,280\,448\left[\frac{1}{8}+\frac{1}{8}\right]} = 0.087\,006\,391$$

C ($n = 6$) vs D ($n = 8$)

$$\sqrt{0.030\,280\,448\left[\frac{1}{8}+\frac{1}{6}\right]} = 0.093\,977\,643$$

Next we calculate the minimum significant difference (MSD) for each pair:

$MSD = q \times SE$

Looking up q for number of samples (k) = 4 and df for the within-sample category = 26 (Table D.10) (interpolating between the values for df = 24 (3.90) and df = 30 (3.85)) we find that q is 3.88. We multiply the standard errors throughout by 3.88, to obtain the MSD:

	A	B	C
B	0.3376	–	–
C	0.3646	0.3646	–
D	0.3376	0.3376	0.3646

Finally, we calculate the differences between the means in the table below and compare them to the equivalent MSD in the table above (means are given in brackets; differences between means that are greater than the MSD are highlighted with an asterisk):

	A (3.2125)	B (2.5875)	C (6.0833)
B (2.5875)	0.6250*	–	–
C (6.0833)	2.8708*	3.4958*	–
D (3.4125)	0.2000	0.8250*	2.6708*

The only comparison not significant is that of lake D compared with lake A.

There are significant differences in the levels of organochlorine residues in gull eggs between the four lakes ($F_{3,26} = 256.07$, $P < 0.01$). Tukey–Kramer multiple comparisons indicate that eggs from lake C have the highest and lake B the lowest residues ($P < 0.05$), but lakes A and D do not differ significantly from each other.

The results could also be presented using a letter code above each bar of a histogram as in Figure 7.2.

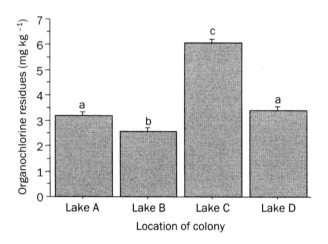

Figure 7.2 *Mean organochlorine residues (mg kg⁻¹) in gull eggs from four colonies. Bars indicate standard errors. Samples labelled with the same letter are not significantly different following Tukey–Kramer's multiple comparisons*

Kruskal–Wallis one-way analysis of variance using ranks

Since parametric ANOVA assumes that the data are normally distributed and the variances are equal, for data that either do not fit these requirements (and cannot be transformed to fit them), or are measured on an ordinal scale, a nonparametric alternative may be required. In the same way that a Mann–Whitney U test provides a nonparametric alternative to the t test for two samples, the Kruskal–Wallis one-way analysis of variance using ranks is a nonparametric alternative used where more than two samples are compared. As in many such nonparametric tests, the calculations are made on the ranks of the data, rather than on the actual data values themselves. Exactly as for the Mann–Whitney U test in Chapter 4, the data are ranked across the whole table (i.e. by placing all the data in ascending order, ignoring the separate samples for the present and giving mean values to tied data – see Box 4.4). The sums of the ranks are then calculated for each sample separately (as for the Mann–Whitney U test). The statistic calculated by the Kruskal–Wallis test is called H and the formula for calculating it is given in Box 7.4. Like the test statistic X^2 (Chapter 6), H approximates to the chi-square distribution. We compare our calculated H value to a chi-square table (an extract of which is given in Table 7.6) at $k - 1$ degrees of

---

Box 7.4 *Formulae for Kruskal–Wallis one-way analysis of variance using ranks*

To compare k samples, calculate the test statistic:

$$H = \left[\frac{12}{n_T(n_T+1)} \times \left(\frac{\left(\sum R_1\right)^2}{n_1} + \frac{\left(\sum R_2\right)^2}{n_2} + \frac{\left(\sum R_3\right)^2}{n_3} + \cdots + \frac{\left(\sum R_k\right)^2}{n_k} \right) \right] - 3(n_T+1)$$

Here k is the number of samples being compared; n_T is the total number of observations across all of the samples; n_1 the number of observations in sample 1, n_2 the number of observations in sample 2, and so on up to sample k; $\sum R_1$ the sum of ranks of observations in sample 1, $\sum R_2$ the sum of ranks of observations in sample 2, and so on up to sample k.

Then compare this H value to chi-square with $k–1$ degrees of freedom (see Table D.8, Appendix D), and reject the null hypothesis if the calculated value is greater than the table value.

Correction for ties

Carry out a parametric ANOVA (see Box 7.1 and Table 7.2) on the ranks of the data, but instead of finding an F value, take $SS_{between}$ and MS_{total} to find the H value:

$$H = \frac{SS_{between}}{MS_{total}}$$

where $SS_{between}$ is the sum of squares for the between variation, and MS_{total} is the SS_{total} divided by the total degrees of freedom.

---

freedom, where k is the number of samples being compared. If the calculated value of H is greater than the table chi-square value, then there are significant differences between the means. The procedure differs when there are three samples all with fewer than five observations (see Siegel and Castellan, 1988). In this case, the statistic does not approximate to the chi-square distribution and alternative tables of H should be consulted (see texts such as Siegel and Castellan 1988).

Although it may not seem so from the formula in Box 7.4, the Kruskal–Wallis test is the same as a one-way parametric ANOVA performed on the ranks of the data (although a different test statistic, an H value rather than an F value, is calculated at the end). The formula in Box 7.4 is much easier to use than a parametric ANOVA of the ranks because it is a simplified version that assumes an absence of ties. However, if there are tied ranks then the test is slightly too conservative (i.e. a type II error is more likely). If your computer program displays H corrected for ties and its associated P value, then it is these you should record. The presence of ties is usually only important if you obtain a marginally non-significant result (i.e. one where the calculated value of H is just less than the tabled value of chi-square with $k – 1$ degrees of freedom). In this instance, it may be worth recalculating it by parametric ANOVA of the ranks and finding H by dividing the $SS_{between}$ by MS_{total} (see the formula at the bottom of Box 7.4). Under these circumstances, you may find that your result is significant after all.

To demonstrate the Kruskal–Wallis test, we will analyse the responses of three samples of people who were asked about their feelings regarding the dangers to the environment of a large oil refinery. One set of people work in the refinery and live nearby, the second set also live nearby but work elsewhere, and the final set neither work at the plant nor live nearby.

WORKED EXAMPLE 7.3A *Kruskal–Wallis analysis of perceptions of the environmental dangers of an oil refinery by people with differing living and working connections to the plant: calculating the* **H** *statistic*

Living nearby and working in the oil refinery (A)		Living nearby and working elsewhere (B)		Living and working elsewhere (C)	
Scores	Rank	Scores	Rank	Scores	Rank
2	7	3	15	3	15
3	15	4	24	4	24
1	2	4	24	2	7
2	7	5	29.5	1	2
2	7	3	15	3	15
1	2	4	24	4	24
3	15	2	7	2	7
4	24	4	24	3	15
3	15	5	29.5	3	15
2	7	4	24	4	24
	$\Sigma R_A = 101$		$\Sigma R_B = 216$		$\Sigma R_C = 148$

The ranked values are placed next to the original scores. Note that the accuracy of the ranking process can be checked by adding up the sums of the ranks $(\Sigma R_A + \Sigma R_B + \Sigma R_C)$ which should equal:

$$\frac{n_T(n_T + 1)}{2}$$

where n_T is the total number of observations across all samples. So:

$$\Sigma R_T = \Sigma R_A + \Sigma R_B + \Sigma R_C = 101 + 216 + 148 = 465$$

continued . . .

Worked Example 7.3a continued . . .

and:

$$\sum R_{\mathrm{T}} = \frac{n_{\mathrm{T}}\left(n_{\mathrm{T}}+1\right)}{2} = \frac{30 \times 31}{2} = 465$$

Therefore the ranking is correct. Next we calculate H:

$$H = \left[\frac{12}{n_{\mathrm{T}}\left(n_{\mathrm{T}}+1\right)} \times \left(\frac{\left(\sum R_{\mathrm{A}}\right)^2}{n_{\mathrm{A}}} + \frac{\left(\sum R_{\mathrm{B}}\right)^2}{n_{\mathrm{B}}} + \frac{\left(\sum R_{\mathrm{C}}\right)^2}{n_{\mathrm{C}}}\right)\right] - 3\left(n_{\mathrm{T}}-1\right)$$

$$= \left[\frac{12}{30 \times 31} \times \left(\frac{101^2}{10} + \frac{216^2}{10} + \frac{148^2}{10}\right)\right] - 3 \times 31$$

$$= \left(0.012\,903\,225 \times 7876.1\right) - 93 = 8.627$$

This value is then compared to chi-square with 2 degrees of freedom $(k-1)$. The calculated value (8.627) is larger than the table value at $P = 0.05$ $(\chi^2 = 5.991)$ and smaller than the table value at $P = 0.01$ $(\chi^2 = 9.210)$, and is therefore significant at $P < 0.05$.

Table 7.6 *Selected values of chi-square (χ^2). Compare H at df = number of samples minus one. Reject the null hypothesis if the calculated value of H^2 is greater than the table chi-square value. Shading indicates the critical values for the example referred to in the text. A more comprehensive table of chi-square values is given in Table D.8 (Appendix D)*

df	χ^2	
$(n-1)$	**P = 0.05**	P = 0.01
1	**3.841**	6.635
2	5.991	9.210
3	**7.815**	11.345
4	**9.488**	13.277
5	**11.070**	15.086

The respondents are asked to score their attitudes on a five-point scale (where 1 indicates no dangers perceived, and 5 many dangers perceived). The data, ranks and calculation of H are given in Worked Example 7.3a. Ranks are calculated as shown in Box 4.4. Although our example has equal sample sizes, it is not necessary for this test.

Note that the data in Worked Example 7.3a are listed in three separate columns (one for each sample). If you use a computer program to calculate your Kruskal–Wallis tests, this data format will not usually be appropriate. See Table B.6 (Appendix B) for details of data entry for statistical analysis using computers.

From the data in Worked Example 7.3a, the H statistic is calculated as 8.63. When compared with chi-square tables at $k-1$ (here $3-1=2$) degrees of freedom (see the shaded area of Table 7.6), we find that the probability that the null hypothesis is true is between $P = 0.05$ ($\chi^2 = 5.991$) and $P = 0.01$ ($\chi^2 = 9.210$). Therefore we can reject the null hypothesis and say:

> There is a significant difference in the perceptions of the environmental dangers of a large oil refinery between people with differing connections (living and working) to the plant ($H = 8.627$, df $= 2$, $P < 0.05$).

There were a large number of ties in our data, therefore our H value is conservative. Since we have rejected the null hypothesis anyway, it does not alter the major outcome. However, if we had taken the tied values into account, H would have been 9.25 (calculated using a parametric ANOVA of the ranks – see the formula at the bottom of Box 7.4), and we would have rejected the null hypothesis with greater confidence ($P < 0.01$).

We can now say there is a difference in the perceptions of the three groups of people towards the dangers to the environment of the oil refinery, but we cannot say definitively where the differences lie. For this we need suitable multiple comparison tests. As with the parametric equivalents, there is a different test for equal (Nemenyi's test) and unequal (Dunn's test) sample sizes. Note that most computer programs do not calculate either Nemenyi or Dunn's tests.

Multiple comparison tests with equal sample sizes: the Nemenyi test

Where sample sizes are equal, the Nemenyi test is used to compare each possible pair of rank means. A rank mean is the sum of the ranks divided by the sample size (i.e. for sample A, the rank mean of A is $\Sigma R_A / n_A$). First, the MSD between rank means is calculated (see Box 7.5). This MSD is then compared to the absolute differences

---

Box 7.5 Formula for the Nemenyi multiple comparison test for use after Kruskal–Wallis analysis of variance using ranks (where sample sizes are equal)

The minimum significant difference is:

$$\text{MSD} = q \times SE$$

where the standard error is:

$$SE = \sqrt{\frac{k(n_T + 1)}{12}}$$

and:

q is the value obtained from the q distribution table using the number of samples involved (k) and degrees of freedom of infinity;

k is the number of samples being compared;

n_T is the total number of observations across the analysis.

(i.e. the differences ignoring any negative signs) between pairs of rank means. If an absolute difference is greater than the MSD, then the difference is significant. As in the Tukey test, this test also uses q and a standard error term, but this time q is consulted at k (the number of samples) and infinite degrees of freedom (given the symbol ∞). Using the rank means of the perceptions of the environmental dangers of a large oil refinery among people with different living and working connections to the plant, the Nemenyi test is calculated in Worked Example 7.3b.

WORKED EXAMPLE 7.3B *Kruskal–Wallis analysis of perceptions of the environmental dangers of an oil refinery by people with differing living and working connections to the plant: calculating the Nemenyi test*

For multiple comparisons on our example data from Worked Example 7.3a:

$$SE = \sqrt{\frac{k(n_T + 1)}{12}} = \sqrt{\frac{3 \times 31}{12}} = 2.783\,882\,181$$

From Table 7.7, q is 3.31, and:

$$MSD = q \times SE = 3.31 \times 2.783\,882\,181 = 9.215$$

This value is compared to the absolute difference between the rank means, given in the table below (rank means are given in brackets; differences between rank means that are greater than the MSD are highlighted with an asterisk):

	Living close and working in refinery, A (10.1)	Living close and working elsewhere, B (21.6)
Living close and working elsewhere, B (21.6)	11.5 *	–
Living and working elsewhere, C (14.8)	4.7	6.8

Table 7.7 *Selected values of q for the Nemenyi test. Shading indicates the critical values for the example referred to in the text. A more comprehensive table of q values is given in Table D.10 (Appendix D)*

		Number of samples being compared (k)		
		3	4	5
Degrees of freedom	∞	3.31	3.63	3.86

From Worked Example 7.3b, the minimum significant difference is 9.215. The only case where the difference between the rank means is greater than the MSD is the comparison of people living close to and working at the plant, with those living close to but working elsewhere:

Living close and working elsewhere > Living close and working in refinery.

However, 'Living and working elsewhere' lies somewhere between the above two; the multiple comparison test is not able to place it. This ambiguity occurs because multiple comparisons are not powerful tests, tending to increase the chances of type II errors (i.e. accepting the null hypothesis when in fact it is false): a further survey (perhaps incorporating larger samples) would be necessary to gain a better picture of where the scores from that group of people lie. We could describe this result in the following way:

> There was an overall significant difference in perception of the environmental dangers of the oil refinery for people with different living and working connections to the refinery ($H = 8.63$, df = 2, $P < 0.05$). Using Nemenyi multiple comparison tests, those living near to and working elsewhere had significantly higher perception scores than those living near and working at the refinery ($P < 0.05$).

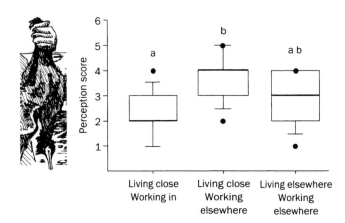

The medians and quartiles (see Boxes 2.2 and 2.7), along with the multiple comparison results, could be displayed in an annotated box and whisker plot as in Figure 7.3.

Figure 7.3 *Median perception scores of the environmental dangers of a large oil refinery by people with differing work and living connections to the plant (n = 10 for each group). Scores range from 1 (no perceived dangers) to 5 (many perceived dangers). Boxes labelled with the same letter are not significantly different following the Nemenyi test for multiple comparisons*

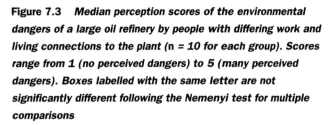

Multiple comparison tests with unequal sample sizes: Dunn's test

Where multiple comparisons are required, but the sample sizes are not equal, Dunn's method is appropriate. Here (like the Tukey– Kramer method for parametric ANOVA) a separate MSD must be calculated for each pair of samples. The formula, given in Box 7.6, is similar to that of the Nemenyi test in Box 7.5, but a different critical value is used (Q, an extract of a table of Q values is shown in Table 7.8). As an example, data on stock damage in commercial forests from three regions of Britain after a series of winter storms are given in Worked Example 7.4. The data are the percentage of the stock from each of 20 forests (8 in region A, 5 in region B and 7 in region C) which show substantial foliar damage (where over 10% of their foliage, relative to a perfect tree, is lost). The overall H value is significant, and

Dunn's multiple comparisons demonstrate where the differences lie, thus we can say that:

> There is a significant difference between the percentage of stock showing foliar damage in three regions ($H = 13.913$, df = 2, $P < 0.01$, $n_A = 8$, $n_B = 5$, $n_C = 7$). Dunn's multiple comparison tests show that the mean rank of region A is lower than the mean ranks of either region B or C, and that regions B and C did not differ significantly from each other ($P < 0.05$).

Box 7.6 Formula for Dunn's multiple comparison test for use after Kruskal–Wallis analysis of variance using ranks (where sample sizes are not equal)

The minimum significant difference is:

$$\text{MSD}_{A,B} = Q \times SE_{A,B}$$

where the standard error (SE) is:

$$SE_{A,B} = \sqrt{\frac{n_T(n_T+1)}{12}\left[\frac{1}{n_A}+\frac{1}{n_B}\right]}$$

and:

$\text{MSD}_{A,B}$ is the minimum significant difference between samples A and B;

Q is the value obtained from the Q distribution table (an extract of which is shown in Table 7.8) using the number of samples (k) – note that this is different to the q table used previously;

n_T is the total number of observations across all the samples;

n_A and n_B are the number of observations in samples A and B.

WORKED EXAMPLE 7.4 *Kruskal–Wallis analysis followed by Dunn's test for stock damage to commercial forests in three regions of Britain*

Region A (n = 8)		Region B (n = 5)		Region C (n = 7)	
% stock	*Rank*	*% stock*	*Rank*	*% stock*	*Rank*
72	1	89	12.5	86	10
76	2	85	9	90	14
80	5	92	15	88	11
82	7	95	19	95	19
84	8	89	12.5	95	19
81	6			93	16
78	3			94	17
79	4				
$\Sigma R_A = 36$		$\Sigma R_B = 68$		$\Sigma R_C = 106$	
Mean rank of A = 4.5		Mean rank of B = 13.6		Mean rank of C = 15.143	

continued . . .

Worked Example 7.4 continued . . .

First check the ranking procedure:

$$\Sigma R_T = \Sigma R_A + \Sigma R_B + \Sigma R_C = 36 + 68 + 106 = 210$$

and

$$\Sigma R_T = \frac{n_T(n_T+1)}{2} = \frac{20 \times 21}{2} = 210$$

The ranking is correct. Next, calculate H:

$$H = \left| \frac{12}{n_T(n_T+1)} \times \left(\frac{\left(\Sigma R_A\right)^2}{n_A} + \frac{\left(\Sigma R_B\right)^2}{n_B} + \frac{\left(\Sigma R_C\right)^2}{n_C} \right) \right| - 3(n_T+1)$$

$$= \left[\frac{12}{20 \times 21} \times \left(\frac{36}{8} + \frac{68}{5} + \frac{106}{7} \right) \right] - 3 \times 21$$

$$= (0.028\,571\,428 \times 2\,691.942\,857) - 63 = 13.913$$

From Table D.8, Appendix D, H is significant ($P < 0.01$) at df $= k - 1 = 3 - 1 = 2$. We carry out Dunn's multiple comparisons, first finding the standard error (SE) for each comparison using:

$$SE_{A,B} = \sqrt{\frac{n_T(n_T+1)}{12}\left[\frac{1}{n_A} + \frac{1}{n_B}\right]}$$

	A ($n = 8$)	B ($n = 5$)

B ($n = 5$) $SE_{A,B} = \sqrt{\dfrac{20 \times 21}{12}\left[\dfrac{1}{8} + \dfrac{1}{5}\right]} = 3.372\,684$

C ($n = 7$) $SE_{A,C} = \sqrt{\dfrac{20 \times 21}{12}\left[\dfrac{1}{8} + \dfrac{1}{7}\right]} = 3.061\,862$ $SE_{B,C} = \sqrt{\dfrac{20 \times 21}{12}\left[\dfrac{1}{5} + \dfrac{1}{7}\right]} = 3.464\,102$

Multiplying throughout by $Q = 2.394$ (see the shaded area of Table 7.8), at $P = 0.05$ (where k is 3) to find MSD values:

	A	B
B	8.07	–
C	7.33	8.29

Worked Example 7.4 continued . . .

Differences in rank means between regions (rank means are given in brackets; differences between rank means that are greater than the MSD are highlighted with an asterisk):

	A (4.5)	B (13.6)
B (13.6)	9.1*	–
C (15.1)	10.6*	1.5

Table 7.8 Selected values of Q for Dunn's test. Values are for P = 0.05. Shading indicates the critical values for the example referred to in the text. A more comprehensive table of Q values is given in Table D.11 (Appendix D)

Number of samples (k)			
3	4	5	6
2.394	2.639	2.807	2.936

Parametric two-way analysis of variance

So far we have investigated differences between categories of one independent variable. However, with analysis of variance it is possible to simultaneously examine the differences due to more than one independent variable. For two independent variables we use a two-way analysis of variance. This also allows us to test for the presence of interactions between the variables: an interaction occurs when a combination of variables produces an effect greater (or less) than would have been expected by examining each variable alone. The design of a two-way experiment or survey can be visualised as in Figure 7.4, where the two independent variables form the r rows and c columns of a two-way design, and each cell is one of $r \times c$ samples, each one representing a particular combination of categories of the independent variable. Here, which of the two independent variables is allocated as the row variable and which is the column variable is arbitrary. If there is only one data point in each cell, then we need a different type of analysis (discussed at the end of this section). In this book, we only consider cases for two-way analysis of variance where the sample sizes are equal. If sample sizes are unequal, the computation and significance testing of the ANOVA are more complex – see texts such as Zar (1999) and Sokal and Rohlf (1995).

Note that Figure 7.4 is a convenient way to visualise two-way designs. However, if you use a computer program to calculate your ANOVA tests, this format will not be

Column variable

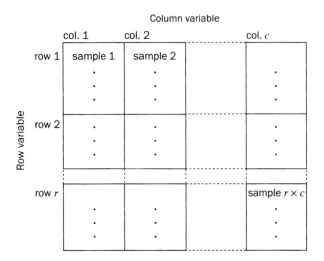

Figure 7.4 Visualising a two-way ANOVA. The dots represent the fact that there are multiple, but equal, sample sizes in each cell, and r is the number of rows; c is the number of columns; r × c is the total number of samples

appropriate. See Table B.6 (Appendix B) for details of data entry for statistical analysis using computers. Also, note that the design in Figure 7.4 does not represent a spatial layout for experimental treatments (e.g. in a field experiment – see Chapter 1 for suitable designs).

As an example of a two-way design, imagine a study of the restoration of derelict land, where the suitability of various substrates for the establishment of wild flower populations was measured by counting the number of species that establish after sowing with a wild flower seed mix. The effect of fertilising the different substrates was also tested by dividing each substrate into those where fertiliser in aqueous solution was applied and those where an equivalent amount of water was applied as a control. Substrate type (which was either brick rubble, colliery spoil or subsoil) was therefore one independent variable, and application (whether fertiliser or control) was the second independent variable. This layout forms $r \times c = 2 \times 3 = 6$ samples: two application types × three substrate types.

First, as for one-way ANOVA, we check for equality of variances by calculating F_{max} for the largest and smallest variances of the $r \times c$ samples. Then, if the F_{max} test shows no significant difference between the largest and smallest variances, we compute F values to obtain the probability of there being no effect due to the: column variable (e.g. substrate); row variable (e.g. application); interaction between the two (e.g. between application and substrate).

The calculations are similar to those in a one-way ANOVA. The total variation is again split into that between categories (i.e. the variation explained by the experiment or survey) and that within samples (the unexplained variation). Therefore:

$$SS_{total} = SS_{between} + SS_{within}$$

However, for two-way ANOVA, the explained variation ($SS_{between}$) contains variation due to two independent variables ($SS_{row\ variable}$ and $SS_{column\ variable}$) and due to the interaction component ($SS_{interaction}$). That is:

$$SS_{between} = SS_{row\ variable} + SS_{column\ variable} + SS_{interaction}$$

It can be seen from the above formula that the more of the variation that is explained by the two independent variables, the less interaction there will be. The total variation breaks down into:

$$SS_{total} = SS_{within} + SS_{row\ variable} + SS_{column\ variable} + SS_{interaction}$$

Details of how to calculate each of the sums of squares are given in Box 7.7. As in one-way ANOVA, the sums of squares and degrees of freedom are entered into an ANOVA table, the calculations for which are shown in Table 7.9.

Box 7.7 *Formulae for the sums of squares for parametric two-way ANOVA*

We find SS_{total} in the same way as for one-way ANOVA (Box 7.1), using all the data combined irrespective of sample:

$$SS_{total} = \sum x_T^2 - \frac{\left(\sum x_T\right)^2}{n_T}$$

where: x_T represents all of the data points irrespective of sample; and n_T is the total sample size.

SS_{within} is found as if we were calculating the SS_{within} for a one-way ANOVA with $r \times c$ separate samples:

$$SS_{within} = SS_1 + SS_2 + SS_3 + \ldots + SS_{r \times c}$$

where SS_1 is the sum of squares for sample 1, SS_2 the sum of squares for sample 2, and so on up to sample $r \times c$ (each sample being a combination of one row and one column variable).

$SS_{row\ variable}$ is found as if we were calculating a one-way ANOVA with r categories, ignoring the different column categories for the present:

$$SS_{row\ variable} = \frac{\left(\sum x_{row\ 1}\right)^2}{n_{row\ 1}} + \frac{\left(\sum x_{row\ 2}\right)^2}{n_{row\ 2}} + \cdots + \frac{\left(\sum x_{row\ r}\right)^2}{n_{row\ r}} - \frac{\left(\sum x_{row\ T}\right)^2}{n_T}$$

Then $SS_{column\ variable}$ is found as if for one way ANOVA on c categories ignoring the row categories:

$$SS_{column\ variable} = \frac{\left(\sum x_{column\ 1}\right)^2}{n_{column\ 1}} + \frac{\left(\sum x_{column\ 2}\right)^2}{n_{column\ 2}} + \cdots + \frac{\left(\sum x_{column\ c}\right)^2}{n_{column\ c}} - \frac{\left(\sum x_T\right)^2}{n_T}$$

Finally, the interaction is calculated as the variation left unaccounted for once all the other components have been calculated (i.e. the total variation minus the other variation components):

$$SS_{interaction} = SS_{total} - SS_{within} - SS_{row\ variable} - SS_{column\ variable}$$

Table 7.9 *Producing a two-way ANOVA table*

	Degrees of freedom (df)	Sums of squares (SS)	Mean squares (MS)	Test statistic (F)	Probability (P): consult table at
Between					
Row variation	$r-1$	$SS_{\text{row variable}}$	$\dfrac{SS_{\text{row variable}}}{df_{\text{row variation}}}$	$\dfrac{MS_{\text{row variable}}}{MS_{\text{within}}}$	$df_{\text{row variation}}$ and df_{within}
Column variation	$c-1$	$SS_{\text{column variable}}$	$\dfrac{SS_{\text{column variable}}}{df_{\text{column variation}}}$	$\dfrac{MS_{\text{column variable}}}{MS_{\text{within}}}$	$df_{\text{column variation}}$ and df_{within}
Interaction	$(r-1)(c-1)$	$SS_{\text{interaction}}$	$\dfrac{SS_{\text{interaction}}}{df_{\text{interaction}}}$	$\dfrac{MS_{\text{interaction}}}{MS_{\text{within}}}$	$df_{\text{interaction}}$ and df_{within}
Within	$rc(n-1)$	SS_{within}	$\dfrac{SS_{\text{within}}}{df_{\text{within}}}$		
Total	n_T-1	SS_{total}			

$MS_{\text{row variable}}$, $MS_{\text{column variable}}$ and $MS_{\text{interaction}}$ are all obtained by dividing the equivalent SS terms by their respective degrees of freedom. The F values for application, substrate and interaction are obtained by dividing their MS term by MS_{within}. The probabilities are obtained from the table of F values (Table D.7, Appendix D). The data and ANOVA calculations for the number of plant species establishing on different substrates and with different applications are given in Worked Example 7.5a.

The three F values obtained from our example are shown in the table at the bottom of Worked Example 7.5a. Each of the three are compared with the table F values at the appropriate degrees of freedom (see Table D.7, Appendix D). For example, the calculated value of F for application is 4.777. This is higher than the table value at df of 1 and 24 for $P = 0.05$ (4.26), but lower than that at $P = 0.01$ (7.82), therefore fertiliser application has a significant effect ($P < 0.05$). The results from Worked Example 7.5a show that both substrate and application have a significant effect (the former being highly significant), with no significant interaction. The fact that there is no interaction indicates that substrate type and the addition of fertiliser have independent effects on the number of species. The results of a two-way ANOVA can be conveniently represented as an interaction line plot (Figure 7.5). The plot shows both the significant effects. One effect is the difference between substrate types, with subsoil (the top line) supporting the highest number of species, followed by brick rubble and then colliery spoil. This order of substrate types is the same whether or not fertiliser is added. The fertiliser has a similar effect on all three substrates, increasing the number of species in about the same proportion for each substrate type (as can be seen by the fact that the lines are nearly parallel). Thus, the effect of fertiliser does not depend on substrate type, but is consistent for each substrate type. When a significant interaction is present, the lines on a two-way line plot are not parallel. Later we explain how to interpret a significant interaction.

WORKED EXAMPLE 7.5A *Two way ANOVA on plant establishment in different substrates with and without fertiliser applications: calculation of the F statistics*

Substrate (columns)

	Brick rubble	Colliery spoil	Subsoil	Application totals
Fertiliser	Sample 1	Sample 2	Sample 3	Fertiliser total
	12	11	16	(n_{Fert} = 15)
	13	10	12	
	10	8	14	Σx_{Fert} = 180
	12	10	15	Σx^2_{Fert} = 2224
	11	12	14	

Applications (rows)

Σx_1 = 58	Σx_2 = 51	Σx_3 = 71
Σx_1^2 = 678	Σx_2^2 = 529	Σx_3^2 = 1017
SS_1 = 5.2	SS_2 = 8.8	SS_3 = 8.8
s_1^2 = 1.3	s_2^2 = 2.2	s_3^2 = 2.2

	Brick rubble	Colliery spoil	Subsoil	
Control	Sample 4	Sample 5	Sample 6	Control total
	12	9	12	($n_{Control}$ = 15)
	10	6	14	
	11	9	14	$\Sigma x_{Control}$ = 163
	12	10	13	$\Sigma x^2_{Control}$ = 1865
	8	8	15	

Σx_4 = 53	Σx_5 = 42	Σx_6 = 68
Σx_4^2 = 573	Σx_5^2 = 362	Σx_6^2 = 930
SS_4 = 11.2	SS_5 = 9.2	SS_6 = 5.2
s_4^2 = 2.8	s_5^2 = 2.3	s_6^2 = 1.3

Substrate totals	Brick rubble total	Colliery spoil total	Subsoil total	Grand total
	(n_{Brick} = 10)	($n_{Colliery}$ = 10)	($n_{Subsoil}$ = 10)	(n_T = 30)
	Σx_{Brick} = 111	$\Sigma x_{Colliery}$ = 93	$\Sigma x_{Subsoil}$ = 139	Σx_T = 343
	Σx^2_{Brick} = 1251	$\Sigma x^2_{Colliery}$ = 891	$\Sigma x^2_{Subsoil}$ = 1947	Σx_T^2 = 4089

First we check the equality of the variances: F_{max} = 2.8 / 1.3 = 2.153; the table value (Table D.9, Appendix D) for F_{max} at number of samples = 6 and sample df = $n - 1 = 4$ is 29.5. The variances are not significantly different ($P > 0.05$). So we proceed with the ANOVA:

continued . . .

Worked Example 7.5a continued . . .

$$SS_{within} = SS_1 + SS_2 + SS_3 + SS_4 + SS_5 + SS_6$$
$$= 5.2 + 8.8 + 8.8 + 11.2 + 9.2 + 5.2$$
$$= 48.4$$

$$SS_{application} = \frac{\left(\sum x_{Fert}\right)^2}{n_{Fert}} + \frac{\left(\sum x_{Control}\right)^2}{n_{Control}} - \frac{\left(\sum x_T\right)^2}{n_T}$$

$$= \frac{(180)^2}{15} + \frac{(163)^2}{15} - \frac{(343)^2}{30}$$

$$= 9.633\,333\,33$$

$$SS_{substrate} = \frac{\left(\sum x_{Brick}\right)^2}{n_{Brick}} + \frac{\left(\sum x_{Colliery}\right)^2}{n_{Colliery}} + \frac{\left(\sum x_{Subsoil}\right)^2}{n_{Subsoil}} - \frac{\left(\sum x_T\right)^2}{n_T}$$

$$= \frac{(111)^2}{10} + \frac{(93)^2}{10} + \frac{(139)^2}{10} - \frac{(343)^2}{30}$$

$$= 107.466\,666\,67$$

$$SS_{total} = \sum x_T^2 - \frac{\left(\sum x_T\right)^2}{n_T}$$

$$= 4089 - \frac{(343)^2}{30}$$

$$= 167.366\,666\,7$$

$$SS_{interaction} = SS_{total} - SS_{within} - SS_{application} - SS_{substrate}$$

$$= 167.366\,667 - 48.4 - 9.633\,333\,3 - 107.466\,666$$

$$= 1.866\,67$$

We can now fill in the ANOVA table:

		df	SS	MS	F	P
	Application	1	9.633 333	9.6333	4.777	< 0.05
Between	Substrate	2	107.466 667	53.7333	26.644	< 0.01
	Interaction	2	1.866 667	0.9333	0.463	> 0.05
Within		24	48.400 000	2.0167		
Total		29	167.366 667			

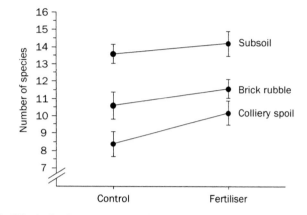

Figure 7.5 *Effect of substrate and application of fertiliser on the number of plant species establishing. Bars represent standard errors*

Note that a line drawn on a plot usually indicates a continuous scale on both the x and y axes, and that the line can be used to obtain a value of y for any value of x. Obviously in this case there are no intermediate points on the x axis, and the lines serve only to illustrate a potential interaction.

The ANOVA results tell us that there is a significant effect of application. Inspection of the mean number of species (calculated from the fertiliser total in Worked Example 7.5a, $\bar{x} = 180/15 = 12$; and from the control total, $\bar{x} = 163/15 = 10.9$) tells us that the significant effect is due to there being more species in the fertiliser treatment than in the control. Since there are only two treatments, this result is unambiguous. However, for substrate type we cannot say definitively that Subsoil > Brick rubble > Colliery spoil; we can only say that there is a difference between the means. Multiple comparisons between the means of the three substrate types are necessary to find out where the differences lie. Multiple comparisons are performed using the Tukey test with the same method as given for one-way ANOVA (Box 7.2). Here, because we are comparing the means of the three different substrates, n is 10 for each group and the number of comparisons (k) is 3. The calculations are shown in Worked Example 7.5b. Since all comparisons are significant, we conclude that: Subsoil > Brick rubble > Colliery spoil (at $P = 0.05$). We could report this as follows:

> There was a significant effect on the number of species establishing of both substrate type ($F_{2,24} = 26.64$, $P < 0.01$) and fertiliser ($F_{1,24} = 4.78$, $P < 0.05$), but no interaction ($F_{2,24} = 0.46$, $P > 0.05$). Tukey's multiple comparisons revealed that all three substrate types differed from each other, with subsoil supporting the most species, and colliery spoil supporting the least ($P < 0.05$). The addition of fertiliser increased the number of species on all three substrates.

The rules for multiple comparisons are different if there is a significant interaction (see the example in the next subsection). In general, when there is no significant interaction we carry out the multiple comparisons for each significant variable with more than two means to be compared, ignoring the other variable. So, had we needed

WORKED EXAMPLE 7.5B *Two-way ANOVA on plant establishment in different substrates with and without fertiliser applications: calculation of the minimum significant difference*

The minimum significant difference is:

$$MSD = q \times SE$$

The value of q for three comparisons at degrees of freedom for MS_{within} of 24 is 3.53 (from Table 7.5), and the standard error is:

$$SE = \sqrt{\frac{MS_{within}}{n_{substrate}}} = \sqrt{\frac{2.0167}{10}} = 0.4491$$

So:

$$MSD = 3.53 \times 0.4491 = 1.585$$

The absolute differences between the means are as follows (means are given in brackets; differences between means that are greater than the MSD are highlighted with an asterisk):

	Brick rubble (11.1)	Colliery spoil (9.3)
Colliery spoil (9.3)	1.8*	–
Subsoil (13.9)	2.8*	4.6*

to perform multiple comparison tests on the fertiliser data (i.e. if we had more than two treatments and there was a significant difference found using the analysis of variance), we would repeat the process in Worked Example 7.5b, for fertiliser application, ignoring substrate type.

Interpretation of significant interactions

Having considered how to deal with a two-way ANOVA where the interaction is not significant, we will now examine an example with a significant interaction between the independent variables. If, instead of fertiliser, the previous experiment had compared the effect of applying lime to the three substrate types with a control treatment of water, the ANOVA table could look like Table 7.10.

This time, the effect of substrate type is significant, but application (whether lime or control) is not. There is also a significant interaction between application and substrate, implying that lime does not behave in the same way on all the substrate types. To visualise the interaction, it is useful to produce an interaction line plot, as in Figure 7.6. When an interaction occurs, the lines on an interaction line plot are not

Table 7.10 *ANOVA table of plant establishment in different substrates with and without the application of lime (n = 5 in each sample)*

Source	df	SS	MS	F	P
Between					
Application	1	0.833	0.833	0.221	> 0.05
Substrate	2	39.200	19.600	5.204	< 0.05
Interaction	2	68.267	34.133	9.062	< 0.01
Within	24	90.400	3.767		
Total	29	198.700			

parallel. It can be seen here that while liming appears to marginally decrease the number of species on brick rubble and subsoil, it has the opposite effect on the colliery spoil, increasing the number of species present. The interaction also helps to explain the lack of significance for the liming treatment: on average over the three substrates, lime neither increases nor decreases the number of species. To explain a significant interaction, some background knowledge of the material under study is often required. In this example, it is possible that lime (which is alkaline) has neutralised the acidity of colliery spoil, making it more favourable for plant growth. Conversely, it possibly makes the other substrates more alkaline and thereby marginally less favourable.

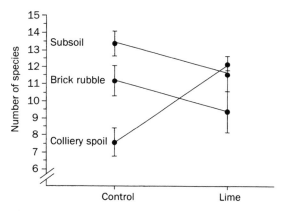

Figure 7.6 *Effect of substrate type and addition of lime on the number of plant species establishing. Bars represent standard errors.*

Although there is a significant difference between the number of species establishing on the different substrate types, the presence of a significant interaction makes interpretation of this effect difficult (and meaningless, according to some statisticians). Consequently, it is more worthwhile to look at the means of each sample separately. The Tukey test for multiple comparisons between all six samples will find where the differences lie, although we must bear in mind that with the large number of tests involved (15), these tend not to be very powerful. The method is the same as before (Box 7.2) and is demonstrated in Worked Example 7.6. Four of the comparisons are significant. Notice that what appeared to be a marginal decrease in number of species with addition of lime on brick rubble and subsoil is not significant, whereas lime has significantly increased the number of species growing on colliery spoil. The means can be placed in order and the differences visualised as:

Colliery Control	Brick Lime	Brick Control	Subsoil Lime	Colliery Lime	Subsoil Control

Here the lines link means that are not significantly different after the Tukey test. Alternatively, the line plot could be annotated with letters as shown in Figure 7.7.

WORKED EXAMPLE 7.6 *Calculation of the minimum significant difference for plant establishment in different substrates with and without the application of lime*

The minimum significant difference is:

$$MSD = q \times SE$$

To compare means when there are $n = 5$ data points in each sample and the MS_{within} is 3.767, the standard error is:

$$SE = \sqrt{\frac{MS_{within}}{n}} = \sqrt{\frac{3.767}{5}} = 0.867\,99$$

From Table D.10 (Appendix D), the value of q with six samples being compared and df = 24, is 4.37. Therefore:

$$MSD = 4.37 \times 0.867\,99 = 3.793$$

The absolute differences between the means are as follows (means are given in brackets; differences between means that are greater than the MSD are highlighted with an asterisk):

			Brick		Colliery spoil		Subsoil	
			control (11.2)	lime (9.4)	control (7.6)	lime (12.2)	control (13.4)	lime (11.6)
Brick	control	(11.2)	–					
	lime	(9.4)	1.8	–				
Colliery spoil	control	(7.6)	3.6	1.8	–			
	lime	(12.2)	1.0	2.8	4.6*	–		
Subsoil	control	(13.4)	2.2	4.0*	5.8*	1.2	–	
	lime	(11.6)	0.4	2.2	4.0*	0.6	1.8	–

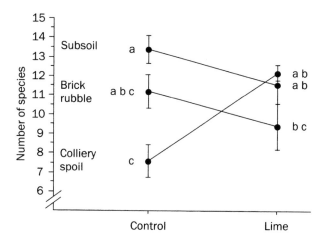

Figure 7.7 *The interaction between substrate type and addition of lime on the number of plant species establishing. Bars represent standard errors. Means labelled with the same letter are not significantly different at P = 0.05 using the Tukey multiple comparison test*

We have seen that the way in which multiple comparisons are calculated for two-way analyses of variance depends on whether there is a significant interaction. Figure 7.8 gives guidelines for deciding which categories or samples to compare.

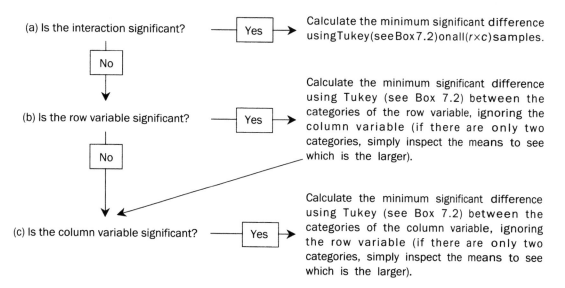

Figure 7.8 *Guidelines for multiple comparison tests in two-way ANOVA*

Two-way ANOVA with single observations in each cell

Two-way analyses of variance can also be performed in cases where there are single observations for each sample (i.e. there is only one data point for any combination of the two independent variables). This experimental design is represented in Figure 7.9. Note that although Figure 7.9 is a convenient way to visualise this two-way design, it is not usually an appropriate way of analysing such designs using computer programs. See Table B.6 (Appendix B) for details of data entry for statistical analysis using computers.

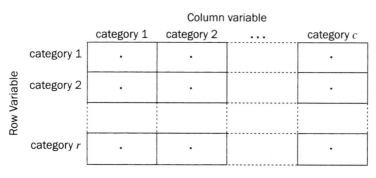

Figure 7.9 *Visualising a two-way ANOVA with single observations in each cell*

Since there is only one data point per sample, it is not possible to calculate the interaction between the variables and we have to assume that there is none. There is no within variation term because there cannot be variation within samples if there is only one data point in each sample. The ANOVA table is therefore simpler, with three types of variation: that due to the column variable; that due to the row variable; and that unaccounted for by either independent variable. This unaccounted-for variation is called the remainder, and its mean square is used as the **denominator** in the calculation of the F value. The degrees of freedom associated with the remainder and with the appropriate variable are used to compare the calculated F value to the table values of F. The rules for calculating the degrees of freedom are given in the ANOVA table at the end of Worked Example 7.7. The formulae for calculating the row and column sums of squares are the same as used previously (Box 7.7), and the $SS_{remainder}$ values are calculated as SS_{total} minus both $SS_{row\ variable}$ and $SS_{column\ variable}$.

To illustrate the technique, consider the amount of water in cubic metres per capita per annum abstracted in 1990 for three purposes (domestic, industrial and agricultural use) in each of six different regions of the world. This is an example where there

cannot be more than one data point in each sample. If we wished to investigate whether there was a difference between regions in the amount of water abstracted, and simultaneously test whether there was a difference between the amount abstracted for each purpose, we would carry out a two-way analysis of variance (see Worked Example 7.7). From this example we can see that the F value for abstraction purpose is 1.843. When we look this up against the degrees of freedom for abstraction purpose of 2 and remainder degrees of

WORKED EXAMPLE 7.7 *Two-way ANOVA on the amount of water abstracted (cubic metres per capita per annum) for different purposes across six regions of the world*

		Abstraction purpose			Totals for region	
		Domestic	Industrial	Agricultural	Σx	Σx^2
Region	Africa	17.08	12.2	214.72	244	46 545.2448
	N America	152.28	710.64	829.08	1692	1215 572.0544
	S America	85.68	109.48	280.84	476	98 198.0384
	Asia	31.56	42.08	452.36	526	207 396.3296
	Europe	94.38	392.04	239.58	726	220 001.5224
	Oceania	580.48	18.14	308.38	907	432 384.3144

	Totals for abstraction purpose			Grand total	
Σx	961.46	1 284.58	2 324.96	Σx	Σx^2
Σx^2	377 682.636	672 939.0676	1 169 475.801	4571	2 220 097.504

First calculate SS_{total}:

$$SS_{total} = \sum x_T^2 - \frac{\left(\sum x_T\right)^2}{n_T} = 2\,220\,097.504 - \frac{(4571)^2}{18} = 1\,059\,317.448$$

Then $SS_{abstraction}$ and SS_{region}:

$$SS_{abstraction} = \frac{\left(\sum x_{dom}\right)^2}{n_{dom}} + \frac{\left(\sum x_{ind}\right)^2}{n_{ind}} + \frac{\left(\sum x_{agr}\right)^2}{n_{agr}} - \frac{\left(\sum x_T\right)^2}{n_T}$$

$$= \frac{(961.46)^2}{6} + \frac{(1284.58)^2}{6} + \frac{(2324.96)^2}{6} - \frac{(4571)^2}{18} = 169\,218.2961$$

$$SS_{region} = \frac{\left(\sum x_{Af}\right)^2}{n_{Af}} + \frac{\left(\sum x_{NA}\right)^2}{n_{NA}} + \frac{\left(\sum x_{SA}\right)^2}{n_{SA}} + \frac{\left(\sum x_{As}\right)^2}{n_{As}} + \frac{\left(\sum x_{Eu}\right)^2}{n_{Eu}} + \frac{\left(\sum x_{Oc}\right)^2}{n_{Oc}} - \frac{\left(\sum x_T\right)^2}{n_T}$$

$$= \frac{(244)^2}{3} + \frac{(1692)^2}{3} + \frac{(476)^2}{3} + \frac{(526)^2}{3} + \frac{(726)^2}{3} + \frac{(907)^2}{3} - \frac{(4571)^2}{18} = 431\,012.2778$$

continued . . .

Worked Example 7.7 continued . . .

Use SS_{total}, $SS_{abstraction}$ and SS_{region} to calculate $SS_{remainder}$:

$$SS_{remainder} = SS_{total} - SS_{abstraction} - SS_{region}$$

$$SS_{remainder} = 1\ 059\ 317.448 - 169\ 218.2961 - 431\ 012.2778 = 459\ 086.8741$$

The ANOVA table can now be completed:

	df			SS	MS	F	P
Abstraction	$(c-1)$	=	2	169 218.2961	84 609.148 05	1.843	> 0.05
Region	$(r-1)$	=	5	431 012.2778	86 202.455 56	1.878	> 0.05
Remainder	$(r-1)(c-1)$	=	10	459 086.8741	45 908.687 41		
Total	n_T	=	17	1 059 317.448			

freedom of 10, the table F value is 4.10 (Table D.7, Appendix D). Since our figure is lower than the table value, there is no significant effect of abstraction purpose. Similarly, we have no significant effect of region since our value of F for region is 1.878, while the table value for region degrees of freedom of 5 and remainder degrees of freedom of 10 is 3.33. Therefore, we can say:

> There is no significant difference in the quantities of water abstracted for each use ($F_{2,10} = 1.843$, $P > 0.05$) and there are no significant differences between the quantity of water abstracted by the regions examined ($F_{5,10} = 1.878$, $P > 0.05$).

If there had been a significant difference in one or both variables then we could have used the Tukey test to find which categories differed from each other, starting at step (b) of Figure 7.8.

Another use of this type of design is to analyse experiments set up in blocks. In Chapter 1 we discussed how an experimental plot could be divided into random blocks in which to lay out the treatments, and how a block could also represent other ways of splitting up the data collection, such as the day on which different measurements were taken. We could consider the blocks to be rows of the ANOVA illustrated in Figure 7.9, and thus also measure variation due to block. For this reason analyses of variance with single observations in each cell are sometimes called randomised block ANOVAs.

We could also use this design of ANOVA to analyse an experiment or survey where we had taken repeated measurements on the same individual sampling units after a number of treatments, or repeated measurements in time or space. This same analysis

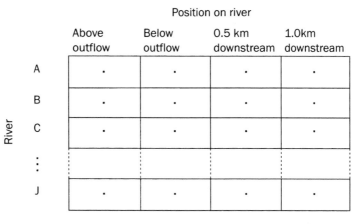

Figure 7.10 *Visualising a repeated measures ANOVA. Single data points are illustrated by the single dots in each cell.*

would then be called a **repeated measures** design (with the individual entered as a row), and the two-way ANOVA would measure whether there was a difference between treatments as well as whether there was variability between individuals. Let us consider our earlier matched-pair example from Chapter 4 (see Worked Example 4.3) where we examined the invertebrate numbers above and below a sewage outflow. Suppose we were also interested in the numbers of invertebrates at two further positions downstream (0.5 km and 1 km below the outflow). This could be analysed using a repeated measures analysis of variance, using the design shown in Figure 7.10. Note that Figure 7.10 is a convenient way to visualise this two-way design, and is an appropriate way of analysing such designs using many computer programs. See Table B.7 (Appendix B) for details of data entry for statistical analysis using computers.

Note that if pairs of data were in each row (or block, or were recorded for each individual) then the research design would be exactly the same as for the paired tests in Chapter 4. In fact, if we computed an ANOVA on the paired *t*-test example in Chapter 4 (see Worked Example 4.3) we would get exactly the same probability for the significance of the column variable (above and below the outflow), and the square of the *t* value would equal *F*. However, the advantage of ANOVA is that we would gain information as to the variability in the individuals (or blocks) under study (in the example of Chapter 4 we would have found whether there was a significant difference between rivers).

Two-way analysis of variance using ranks

Where we have more than one independent variable and data which are on an ordinal scale, or which are not normally distributed (and unsuitable for transformation), then a nonparametric two-way analysis of variance using ranks is appropriate. For this test, the layout of the rows and columns is the same as illustrated for parametric two-way ANOVA (Figure 7.4), and equal sample sizes in each combination of the two variables are necessary to compute the test.

The earlier example for the one-way analysis of variance by ranks was a survey of the perceptions of three groups of people to the environmental dangers of a large oil refinery (see Worked Example 7.3a). The three groups comprised people living close and working within the plant, living close and working elsewhere, and living elsewhere and working elsewhere. Suppose there had been sufficient people working within the plant but living elsewhere to add to the survey. This survey design would then lend itself to a two-way analysis, with two independent variables: where people live (far or close to the plant) and where they work (within the plant or not). Two-way analysis of variance enables us to test whether these two variables affect people's perceptions of the environmental dangers of the refinery. Additionally, the analysis allows us to test for an interaction between the two independent variables. The concept of interaction is the same as for two-way parametric ANOVA. In this example there would be an interaction if working or not at the plant had different effects on the perceptions of people living close to the plant than on those of people living further away.

The test itself is an extension of Kruskal–Wallis one-way analysis of variance using ranks, although it seems more like carrying out a parametric ANOVA (in fact it is the equivalent to carrying out the correction for ties mentioned for one-way Kruskal–Wallis: see the bottom of Box 7.4). We first rank the data across the entire data set and then use the ranks as if they were measured variables in a parametric ANOVA to compute sums of squares for each of the total variation, the row variable, the column variable and the interaction, using the formulae given earlier (see Box 7.7). The difference between the parametric ANOVA and the analysis of variance on the ranks comes at the final step: instead of computing F, we compute H values, which are the relevant sums of squares divided by the total mean square (see Box 7.8).

We then test the three null hypotheses for a two-way analysis of variance: that there is no effect of the row variable; that there is no effect of the column variable; that there is no interaction between the two independent variables. We do this by comparing the calculated values of $H_{row\ variable}$, $H_{column\ variable}$ and $H_{interaction}$, given in Box 7.8, to tabulated values of chi-square (Table D.8, Appendix D) at the appropriate degrees of freedom (see the bottom of Box 7.8): if any exceeds the table values, we reject that particular null hypothesis. Since the method for calculating the sums of squares is exactly the same as for parametric two-way ANOVA, the general formulae are not repeated here. However, in Worked Example 7.8, the symbol R is used instead of x in the formulae to signify that the ranks of the data are used. Note that in Box 7.8 the only mean square value calculated is MS_{total}, which is the SS_{total} divided by the total degrees of freedom.

Box 7.8 *Formulae for the H statistics for two-way analysis of variance using ranks*

First convert the data to ranks, treating all the data as one sample. Then, using the ranked values, calculate $SS_{\text{row variable}}$, $SS_{\text{column variable}}$, $SS_{\text{interaction}}$ and SS_{total} using the formulae in Box 7.7. Three H values are required, one for each of the variables (row and column) and one for the interaction. Their calculation and that of their associated degrees of freedom are given in the following table, where r is the number of rows and c is the number of columns.

	df	MS	H	P: Look up H against chi-square at:
Row variable	$r-1$		$\dfrac{SS_{\text{row variable}}}{MS_{\text{total}}}$	row df
Column variable	$c-1$		$\dfrac{SS_{\text{column variable}}}{MS_{\text{total}}}$	column df
Interaction	$(r-1)(c-1)$		$\dfrac{SS_{\text{interaction}}}{MS_{\text{total}}}$	interaction df
Total	$n_T - 1$	$SS_{\text{total}} / df_{\text{total}}$		

Using our oil refinery example, if we have 'Living' as the column variable and 'Working' as the row variable, there are two rows and two columns, giving us four samples (Worked Example 7.8). We can then identify whether where people work or live in relation to the plant has a significant effect on their perceptions of the environmental dangers posed by the refinery. Note that the data in the analysis of variance table in Worked Example 7.8 are listed in a convenient way to visualise the calculations. This data format will not be appropriate if you use a computer program to calculate your ANOVA tests. See Appendix B (Table B.6) for details of data entry for statistical analysis using computers.

Unlike parametric ANOVA, we look up each test statistic (H) against chi-square values at only one value of degrees of freedom (i.e. we do not use the within degrees of freedom). For each test shown here (the effects of working location, living location and the interaction), the degrees of freedom are 1. From these data, we find that whether or not the respondent works at the refinery is the significant factor affecting their perception of the environmental danger of the refinery ($P < 0.01$). There is no effect of how close the respondent lives to the plant ($P > 0.05$), and no interaction ($P > 0.05$). These results can be visualised on a two-way plot of the rank means (see Figure 7.11). Inspection of the plot allows us to interpret the significant effect of where the respondents work: those who work at the plant have significantly lower scores.

WORKED EXAMPLE 7.8 *Two way ANOVA using ranks for perceptions of the environmental dangers of an oil refinery by people with differing living and working connections to the plant*

Living (columns)

	Close		*Far*	
	Sample 1		*Sample 2*	
	Score	*Rank*	*Score*	*Rank*
	2	10.5	3	21.5
	3	21.5	4	33
	1	3	1	3
	2	10.5	2	10.5
	2	10.5	2	10.5
	1	3	3	21.5
	3	21.5	1	3
In plant	4	33	4	33
	3	21.5	2	10.5
	2	10.5	3	21.5

$\sum R_{Plant} = 313.5$

$\sum R^2_{Plant} = 6848.25$

$n_{Plant} = 20$

$\sum R_{sample\ 1}$	=	145.5	$\sum R_{sample\ 2}$	=	168
$\sum R^2_{sample\ 1}$	=	2934.75	$\sum R^2_{sample\ 2}$	=	3913.5
$SS_{sample\ 1}$	=	817.725	$SS_{sample\ 2}$	=	1091.1

Working (rows)

	Sample 3		*Sample 4*	
	Score	*Rank*	*Score*	*Rank*
	3	21.5	3	21.5
	4	33	4	33
	4	33	2	10.5
	5	39.5	1	3
	3	21.5	3	21.5
	4	33	4	33
	2	10.5	2	10.5
Elsewhere	4	33	3	21.5
	5	39.5	3	21.5
	4	33	4	33

$\sum R_{Not} = 506.5$

$\sum R^2_{Not} = 14945.75$

$n_{Not} = 20$

$\sum R_{sample\ 3}$	=	297.5	$\sum R_{sample\ 4}$	=	209
$\sum R^2_{sample\ 3}$	=	9600.25	$\sum R^2_{sample\ 4}$	=	5345.5
$SS_{sample\ 3}$	=	749.625	$SS_{sample\ 4}$	=	977.4

Total

$\sum R_{Close}$	=	443	$\sum R_{Far}$	=	377	$\sum R_T$	=	820
$\sum R^2_{Close}$	=	12 535	$\sum R^2_{Far}$	=	2959	$\sum R^2_T$	=	21 794
n_{Close}	=	20	n_{Far}	=	20	n_T	=	40

continued . . .

Worked Example 7.8 continued . . .

where $\sum R$ is the sum of the ranks and n is the number of observations.

Now we can calculate the sums of squares:

$$SS_{total} = \sum R_T^2 - \frac{\left(\sum R_T\right)^2}{n_T} = 21\,794 - \frac{(820)^2}{40} = 4984$$

$$SS_{living} = \frac{\left(\sum R_{Close}\right)^2}{n_{Close}} + \frac{\left(\sum R_{Far}\right)^2}{n_{Far}} - \frac{\left(\sum R_T\right)^2}{n_T} = \frac{(443)^2}{20} + \frac{(377)^2}{20} - \frac{(820)^2}{40} = 108.9$$

$$SS_{working} = \frac{\left(\sum R_{Plant}\right)^2}{n_{Plant}} + \frac{\left(\sum R_{Not}\right)^2}{n_{Not}} - \frac{\left(\sum R_T\right)^2}{n_T} = \frac{(313.5)^2}{20} + \frac{(506.5)^2}{20} - \frac{(820)^2}{40} = 931.225$$

$$SS_{within} = SS_{sample\,1} + SS_{sample\,2} + SS_{sample\,3} + SS_{sample\,4} = 817.725 + 1091.1 + 749.625 + 977.4 = 3635.85$$

$$SS_{interaction} = SS_{total} - SS_{within} - SS_{living} - SS_{working} = 4984 - 3635.85 - 108.9 - 931.225 = 308.025$$

Now we fill out an analysis of variance table to calculate the H values and compare these to the appropriate values of χ^2 (see Table D.8, Appendix D):

	df	SS	MS	H value	P
Working	1	931.225		7.287	< 0.01
Living	1	108.9		0.852	> 0.05
Interaction	1	308.025		2.410	> 0.05
Within	36	3635.85			
Total	39	4984	127.794 871 8		

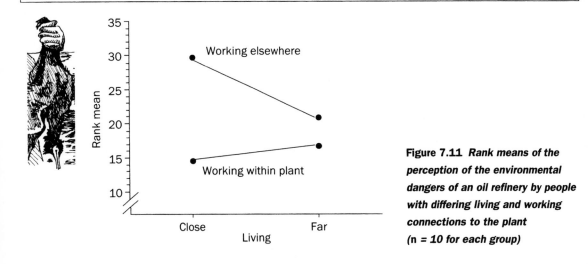

Figure 7.11 Rank means of the perception of the environmental dangers of an oil refinery by people with differing living and working connections to the plant (n = 10 for each group)

From this analysis we can say:

> Those working at the plant perceive the refinery to pose less danger to the environment than those who work elsewhere ($H = 7.287$, df = 1, $P < 0.05$), while those living near to the plant do not differ significantly in perception from those who live further away ($H = 0.852$, df = 1, $P > 0.05$). There is no interaction between the distance respondents live from the plant and whether they work in the plant or elsewhere ($H = 2.410$, df = 1, $P > 0.05$).

Since there are only two alternatives for the significant effect (i.e. working within the plant or elsewhere), inspection of the graph is sufficient to determine how the rank means differ (in this case, those who work at the plant have lower scores than those who work elsewhere). With three or more categories for either independent variable, or if there were a significant interaction, we may wish to carry out multiple comparisons. The Nemenyi test can be used for this (as performed after a significant one-way analysis of variance using ranks: see Box 7.5). To decide exactly which comparisons to perform, follow the flowchart in Figure 7.8: briefly, if there is a significant interaction, irrespective of whether the two independent variables are significant, then compare between all r × c samples; if there is no significant interaction, compare between samples for any significant independent variable.

It should be mentioned here that many introductory statistics textbooks do not cover two-way analysis of variance using ranks. This is because there is some debate about how well the test performs (see Zar, 1999). However, several authors (e.g. Meddis, 1984; Barnard et al., 1993; and Sokal and Rohlf, 1995) do support it, especially when both independent variables are fixed.

Friedman's matched group analysis of variance using ranks

In the same way as for parametric ANOVA, there is a special case of two-way analysis of variance for nonparametric data where there is only one datum in each cell. **Friedman's matched group analysis of variance using ranks** is the appropriate test for this situation. The test can be viewed as an extension to the Wilcoxon matched-pairs test encountered in Chapter 4, where there are more than two samples taken from matched groups. This test is also known under several other names: two-way analysis of variance using ranks (although it is actually a special case of two-way analysis of variance); or as a repeated measures analysis of

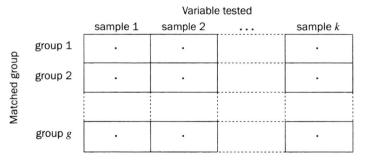

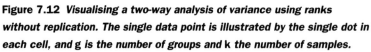

Figure 7.12 *Visualising a two-way analysis of variance using ranks without replication. The single data point is illustrated by the single dot in each cell, and g is the number of groups and k the number of samples.*

variance using ranks; or as a nonparametric randomised block analysis of variance (because it can be used on randomised blocks in the same way as can the parametric two-way ANOVA with single observations in each cell). However, unlike the parametric equivalent, Friedman's method only tests for significance of the column variable; no measure of variation of the matching (row) variable is obtained. Figure 7.12 shows the design of the analysis.

Friedman's test uses ranks of data rather than the actual values, but here data are not ranked across the whole table (as they were for the Kruskal–Wallis test). Because the data are in matched groups, we rank each group separately in turn (i.e. each row of Figure 7.12) across the samples. Following ranking, the sum of ranks for each sample (column) is calculated and the test statistic computed using the formula shown in Box 7.9.

Box 7.9 Formulae for the F_r statistic for Friedman's matched group analysis of variance using ranks

Firstly each group is ranked separately from the others across the samples. The accuracy of the ranking may be checked by adding up the sums of ranks, which should equal:

$$\frac{gk(k+1)}{2}$$

where g is the number of groups of data and k is the number of samples.

The statistic F_r is now calculated as:

$$F_r = \left[\frac{12}{gk(k+1)} \times \left(\left(\sum R_1 \right)^2 + \left(\sum R_2 \right)^2 + \left(\sum R_3 \right)^2 + \cdots + \left(\sum R_k \right)^2 \right) \right] - 3g(k+1)$$

where $\sum R_1$ is the sum of ranks of sample 1, $\sum R_2$ the sum of ranks for sample 2, and so on up to sample k.

If k is large (over 5) or g is large (> 13 for $k = 3$, > 8 for $k = 4$, or > 5 for $k = 5$), F_r is compared to the chi-square distribution with $k - 1$ degrees of freedom. Where g is small, the probability should be obtained from a table of critical values (Table D12, Appendix D).

To illustrate Friedman's method, let us look at fish stocks in lakes within several regions of Scandinavia adversely affected by acid rain. We wish to investigate whether three fish species differ in the extent to which they have been affected. In each of ten regions we assess the condition of the stock of brown trout, perch and Arctic chub to obtain the data shown in Worked Example 7.9a. There are three matched data points for each region. Friedman's test has one independent variable (fish species) and a series of matched groups (the regions). The status of the three species within each region was scored on the following scale:

1 less than 25% of stocks lost or affected, with more affected than lost;
2 less than 25% of stocks lost or affected, with more lost than affected;
3 25–75% of stocks lost or affected, with more affected than lost;
4 25–75% of stocks lost or affected, with more lost than affected;
5 over 75% of stocks lost or affected, with more affected than lost;
6 over 75% of stocks lost or affected, with more lost than affected.

The data in Worked Example 7.9a are listed in three columns (representing the three fish species). If you use a computer program to calculate your ANOVA tests, this data format will be appropriate. See Table B.7 (Appendix B) for details of data entry for statistical analysis using computers.

Note in Worked Example 7.9a that the ranking is done across the rows of the table (i.e. within each group). So for row 1 (region 1) the scores 3, 3, 5 are ranked as 1.5, 1.5, 3. We can see that F_r is 7.8 and we now need to decide whether this is significant. For low numbers of groups (g) and samples (k), we look up F_r in a special table (an extract from which is shown in Table 7.11). When the number of samples exceeds 5 or the number of groups is large (above 13 for $k = 3$; above 8 for $k = 4$; above 5 for $k = 5$), then F_r approximates to chi-square with $k - 1$ degrees of freedom (see Table D.8, Appendix D). Consulting the shaded area of Table 7.11 at $k = 3$ and $g = 10$, we find that our calculated value of F_r (7.8) lies between that at $P < 0.05$ (6.2) and that at $P < 0.01$ (9.6). Therefore the probability of our F_r value occurring by chance is less than 0.05, but greater than 0.01. When there are tied values across the groups, the P value is slightly inaccurate, increasing the likelihood of falsely accepting the null hypothesis (type II error). The more tied ranks there are, the more inaccurate the calculation of the probability. A correction factor can be applied – see texts such as Zar (1999) or Siegel and Castellan (1988). If your computer program gives you corrected values for ties, then record the corrected P value. If you are performing calculations by hand, then this inaccuracy is a problem only if you have a marginally non-significant result (just greater than 0.05). If the result is significant, or is much greater than 0.05, it will remain so even if corrected for ties.

Brown trout appear to be the worst affected, followed by perch and then Arctic chub. As with other analyses of variance tests, we now need a multiple comparison test to determine whether these perceived differences are significant. The test we will use here is the Nemenyi test for matched groups (see Box 7.10) and the calculations for our data are shown in Worked Example 7.9b. We find that the only significant difference is between the trout and perch stocks, with trout being the worse affected. We can therefore state the following:

> There is a significant difference in fish stock damage between the three species in lakes affected by acid rain (F_r matched by region = 7.8, number of regions = 10, $P < 0.05$). Following Nemenyi tests for matched groups, trout are more adversely affected than perch ($P < 0.05$), but Arctic chub did not differ from either trout or perch ($P > 0.05$).

WORKED EXAMPLE 7.9A ***Friedman's matched group analysis of variance using ranks on species differences in fish stock condition in lakes affected by acid rain: calculating the*** $\mathbf{F_r}$ ***statistic***

Categories (k) Fish

Groups (g) Region		Brown trout *Score*	Brown trout *Rank*	Perch *Score*	Perch *Rank*	Arctic chub *Score*	Arctic chub *Rank*
	1	3	1.5	3	1.5	5	3
	2	6	2.5	1	1	6	2.5
	3	3	2	1	1	5	3
	4	3	2.5	1	1	3	2.5
	5	3	2.5	1	1	3	2.5
	6	3	2	3	2	3	2
	7	5	3	1	1	3	2
	8	6	3	4	2	1	1
	9	6	3	4	1.5	4	1.5
	10	4	3	1	1	3	2

$\Sigma R_{\text{trout}} = 25 \qquad \Sigma R_{\text{perch}} = 13 \qquad \Sigma R_{\text{chub}} = 22$

As a check on the ranking procedure, the sum of ranks (across the fish species for each region/group):

$$\sum R_{\text{trout}} + \sum R_{\text{perch}} + \sum R_{\text{chub}} = 25 + 13 + 22 = 60$$

and:

$$\frac{gk(k+1)}{2} = \frac{10 \times 3 \times 4}{2} = 60$$

Therefore, the ranking has been checked and we now calculate the test statistic (F_r) as:

$$F_r = \left[\frac{12}{gk(k+1)} \times \left(R_{\text{trout}}^2 + R_{\text{perch}}^2 + R_{\text{chub}}^2\right)\right] - 3g(k+1)$$

$$= \left[\frac{12}{10 \times 3 \times 4} \times \left(25^2 + 13^2 + 22^2\right)\right] - 3 \times 10 \times 4 = (0.1 \times 1278) - 120 = 7.8$$

This calculated value of F_r is checked against the table value (an extract from an F_r table is shown in Table 7.11) against the number of groups ($g = 10$) and number of samples ($k = 3$). It can be seen that the calculated value ($F_r = 7.8$) is higher than the table value at $P = 0.5$ (6.200) but lower than that at $P = 0.01$ (9.600).

Table 7.11 *Selected values of* F_r *for Friedman's matched group analysis of variance using ranks. Shading indicates the critical values for the example referred to in the text. Reject the null hypothesis if the calculated value of* F_r *is higher than the table value. The upper table values (in bold) are for* P = 0.05, *while the lower values are for* P = 0.01. *A comprehensive table of* F_r *values is given in Table D.12 (Appendix D)*

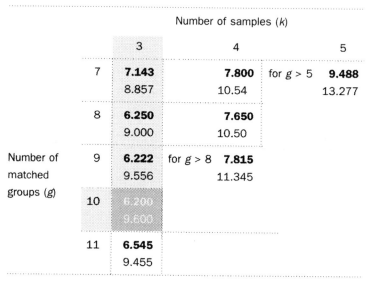

		Number of samples (k)		
		3	4	5
	7	**7.143**	**7.800** for g > 5	**9.488**
		8.857	10.54	13.277
	8	**6.250**	**7.650**	
		9.000	10.50	
Number of	9	**6.222** for g > 8	**7.815**	
matched		9.556	11.345	
groups (g)	10	6.200		
		9.600		
	11	**6.545**		
		9.455		

Box 7.10 Formula for the Nemenyi multiple comparison test for use after Friedman's matched group analysis of variance using ranks

The minimum significant difference is calculated as:

$$MSD = q \times SE$$

where the standard error is:

$$SE = \sqrt{\frac{k(k+1)}{12g}}$$

and:

q is the value from the q distribution table (Table D.10, Appendix D) using the number of samples (k) and degrees of freedom of infinity;

k is the number of samples;

g is the number of groups.

WORKED EXAMPLE 7.9B *Friedman's matched group analysis of variance using ranks on species differences in fish stock condition in lakes affected by acid rain: calculating the minimum significant difference*

Substituting the number of groups and number of samples from Worked Example 7.9a into the formula from Box 7.10, the standard error is:

$$SE = \sqrt{\frac{k(k+1)}{12g}} = \sqrt{\frac{3 \times 4}{12 \times 10}} = 0.316\,23$$

With $k = 3$ and df $= \infty$, $q = 3.31$ (Table D.10, Appendix D), so the minimum significant difference is calculated as:

$$MSD = q \times SE = 3.31 \times 0.316\,23 = 1.014$$

Below are the absolute differences (i.e. ignoring the sign) between pairs of rank means (rank means are given in brackets; differences between means that are greater than the MSD are highlighted with an asterisk):

	Brown trout (2.5)	Perch (1.3)
Perch (1.3)	1.2*	–
Arctic chub (2.2)	0.3	0.9

Testing specific hypotheses

Multiple comparisons are often not very revealing because of their lack of power in detecting differences between means or rank means, especially when several are being compared (e.g. over 20 comparisons). Far more powerful tests can be carried out using specific tests on analysis of variance by ranks (for both one- and two-way tests) if you have specific alternative hypotheses set up in advance of the experiment or survey. For parametric tests, see Zar (1999) and Sokal and Rohlf (1995), and for nonparametric tests see Meddis (1984) for details and Barnard *et al.* (1993) for a simplified brief description. See Chapter 3 for a discussion of specific alternative hypotheses.

Other models of analysis of variance

There are a great many variations on the theme of analysis of variance which are beyond the scope of this book. In this section, the main ones are briefly introduced – see Zar (1999), Sokal and Rohlf (1995) and Manly (1994) for further details.

More than two independent variables

We have considered the case where there are two independent variables. Analysis of variance can be extended to any number of independent variables: for three independent variables three-way analysis of variance is used, for four independent variables four-way analysis of variance, and so on. The greater the complexity of the analysis, the greater the complexity of the experimental or survey design, the larger the sample sizes needed and the more difficult the interpretation (although this is somewhat counterbalanced by the economy of testing several variables at once, and the additional information obtained). The interactions are particularly complex: for three way analysis of variance on variables A, B and C, there are four interactions: $A \times B$, $B \times C$, $A \times C$ and $A \times B \times C$. The number of interactions rapidly increases with the number of independent variables (e.g. there are 11 interactions in a four-way analysis of variance).

Analysing Latin square designs

In some manipulative experiments, plots are laid out in a Latin square design (see Figure 1.3d), where each row and column of an experimental layout contains one replicate of each treatment (in a random order). This type of sophisticated design allows not only the effects of any treatments to be tested, but also the influence of the position (usually to ensure that there is no bias in the design). Such experiments are most common in agricultural systems, and details are given in texts such as Mead *et al.* (1993) and Gomez and Gomez (1984).

Nested analysis of variance

Another type of analysis of variance is used when multiple readings are taken from each of several individuals to take into account the variation within individuals (e.g. measuring the area of four leaves from each of three plants within each of three categories or treatments). This is a nested (or hierarchical) design. The structure of the nested analysis is visualised in Figure 7.13 (the three dots in each cell represent the three leaves sampled from each plant). Note that different plants are used in each category. A nested analysis of variance identifies variation within plants, variation between plants and variation between categories. This design should not be confused with that of a two-way analysis of variance. If plants 1, 2 and 3 each had treatments A, B and C applied to different parts of them, then this would be a two-way design (illustrated in Figure 7.14), the important distinction here being that the same plants are used in each treatment.

For details of analysis using nested designs see texts such as Sokal and Rohlf (1995). Note that neither Figure 7.13 nor 7.14 is intended to represent the actual

Category or treatment

A			B			C		
Plant 1	Plant 2	Plant 3	Plant 4	Plant 5	Plant 6	Plant 7	Plant 8	Plant 9
⋮	⋮	⋮	⋮	⋮	⋮	⋮	⋮	⋮

Figure 7.13 Visualising a nested ANOVA. Each dot represents a data point.

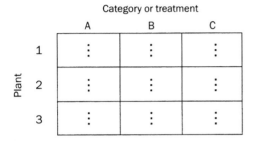

Figure 7.14 Visualising a two-way ANOVA. Each dot represents a data point.

positions of plants or treatments. In the field, the plants and treatments would be selected and laid out in, for example, a randomised block design (see Figure 1.3c). It is also possible to examine data from the design illustrated in Figure 7.13 by taking the mean value for each individual and analysing these using a one-way analysis of variance. However, this method would lose some of the information you have regarding the variation within individuals.

Analysis of covariance

So far, we have considered analyses of one or more independent variables set up as categories or treatments (e.g. insulation type in the example for one-way ANOVA), and measurements of just one dependent variable (energy bill). In the energy bill example (Worked Example 7.1a), only houses of a standard size were sampled. However, if we had been unable to standardise house size, then we could have added house size as another independent variable, recording for each house its size (in cubic metres) as well as the energy bill. A technique called analysis of covariance (ANCOVA), which is a little like a mixture of regression and analysis of variance, could be used on these data. ANCOVA would remove the effect of the independent variable (house size) so that the effect of insulation type on energy bills could be investigated. For further information on this type of analysis, see texts such as Sokal and Rohlf (1995).

Multivariate analysis of variance

If several dependent variables are measured from two or more samples, then a multivariate analysis of variance (MANOVA) is suitable. In the experiment on wild flower establishment on different substrates (used to illustrate two-way ANOVA: Worked Example 7.5a), we could have measured the number of plant species, plant size, etc. and analysed them simultaneously using this technique. For an introduction to MANOVA, see texts such as Zar (1999) and Tabachnick and Fidell (1996).

Summary

By now you should be able to:

- test between the means of three or more normally distributed samples using one-way parametric ANOVA (having first calculated F_{max} to check that the variances are not significantly different), and then carry out multiple comparisons if there is overall significance (using the Tukey test for equal sample sizes or the Tukey–Kramer test for unequal sample sizes);

- test between three or more samples that are measured at least on an ordinal scale using a Kruskal–Wallis test, and then carry out multiple comparisons of rank means if there is overall significance (using the Nemenyi test for equal sample sizes or Dunn's test for unequal sample sizes);

- recognise experimental and survey designs with two independent variables, understand that a two-way analysis is required and appreciate that a third source of variation, the interaction effect, will be present if there is more than one data point in each sample;

- carry out parametric two-way ANOVA on normally distributed data (having first checked for equality of variances using an F_{max} test), interpret the interaction effect, and then carry out multiple comparisons (using the Tukey test) if one or both independent variables and/or the interaction are significant;

- carry out nonparametric two-way analysis of variance using ranks on ordinal or non-normal data, interpret the interaction effect, and then carry out multiple comparisons (using the Nemenyi test) if one or both independent variables and/or the interaction are significant;

- recognise the special case of two-way analysis of variance when there is only one replicate per cell, and use parametric two-way ANOVA without replication (for normally distributed data) or Friedman's matched groups analysis of variance using ranks (for data that are at least ordinal), and use the appropriate multiple comparison test where necessary (the Tukey test for the parametric ANOVA and the Nemenyi matched group test for Friedman's test).

- be aware of other techniques which are appropriate for more complex analyses of variance.

Questions

7.1 NO_2 concentrations (in parts per billion) in the exhaust fumes of different types of vehicle (cars, minibuses and buses) were sampled and the following data were obtained:

Car	Minibus	Bus
69	51	41
62	45	48
73	59	42
70	57	39
68	58	47
68	54	49
64	46	35
71	48	41

(i) Check that the variances of the three samples are not significantly different using an F_{max} test.

(ii) Calculate a one-way ANOVA on the above data and fill out an ANOVA table:

	df	Sums of squares	Mean squares	F value	P
Vehicle type					
Within					
Total					

(iii) Calculate the mean ($\bar{x}$) and standard error (SE) of NO_2 concentration in each vehicle type:

Cars		Minibuses		Buses	
$\bar{x}$:	SE:	$\bar{x}$:	SE:	$\bar{x}$:	SE:

(iv) Using multiple comparison tests where appropriate, what do you conclude from your analysis of these data?

7.2 A local council was considering introducing extensive recycling facilities. As part of a survey investigating the attitudes of the public towards the popularity of this initiative, the effect of age of the respondent was taken into account. Respondents were asked to score their enthusiasm towards recycling on a scale of 1 to 5 (where 5 was very enthusiastic) and their age was classed as: up to 25; from 26 to 50; or over 51 years. The following data were collected:

Up to 25	26–50	Over 51
3	1	4
2	2	3
4	3	3
4	3	5
5	2	3
4	2	1
3	1	4
1	1	4
5	2	5
4	2	4

(i) Carry out Kruskal–Wallis one-way analysis of variance using ranks on the data:

Rank means			Kruskal–Wallis		
Up to 25	26 to 50	Over 51	df:	H:	P:

(ii) Using multiple comparison tests where appropriate, what do you conclude from this analysis?

7.3 The percentages of all forest lost per year in ten randomly selected countries in each of Central America and South East Asia are given below. Half of the countries are island countries, and half are mainland countries. The arcsine-transformed data (to normalise the distributions) are given alongside the original data (the variances of the transformed data have been subjected to an F_{max} test and are not significantly different).

		Region			
		Central America		South East Asia	
		% lost	arcsine- transformed	% lost	arcsine- transformed
	Mainland	3.1	**0.177**	1.0	**0.100**
		3.2	**0.180**	1.5	**0.123**
		2.0	**0.142**	2.5	**0.159**
		2.3	**0.152**	1.7	**0.131**
		2.7	**0.165**	0.2	**0.045**
Location					
	Island	0.1	**0.032**	0.1	**0.032**
		0.6	**0.078**	0.1	**0.032**
		3.8	**0.196**	0.2	**0.045**
		3.0	**0.174**	1.5	**0.123**
		0.4	**0.063**	0.8	**0.090**

(i) Use two-way ANOVA on the transformed data to determine whether forest loss differs depending on: which region the country is in; the location (i.e. whether the country is an island or on the mainland); and whether there is an interaction between region and location.

	df	Sums of squares	Mean squares	F value	P
Region					
Location					
Interaction					
Within					
Total					

(ii) Draw a line plot of the means and standard errors of the transformed data.

(iii) Using multiple comparison tests where appropriate, what do you conclude from this analysis?

7.4 The amount of nitrogen (in kilotonnes per annum) entering UK coastal waters via fresh water inputs (i.e. from rivers) is compared to that entering directly into saline waters from point sources (e.g. pipes):

			Region			
		Atlantic	North Sea (East coast)	North Sea (Channel)	Celtic Sea	Irish Sea
Route	Rivers	28	93	17	30	42
	Direct	10	77	12	11	6

(i) Perform a two-way ANOVA with single observations in each cell (assume there is no interaction between input route and region) to decide whether there is a difference in the amount of nitrogen entering coastal waters: (a) via the two routes; and (b) between regions. Note that the amount of nitrogen entering coastal waters via the two routes could be compared using a paired t test – and yield the same result – but this would not provide information on variation between regions.

	df	SS	MS	F	P
Route					
Region					
Remainder					
Total					

(ii) Using multiple comparison tests where appropriate, what can you conclude from these results?

7.5 An environmental pressure group contacted backbench MPs from two political parties (neither of which were in government), just before and immediately after a general election, and asked them for their views on environmental issues. The responses were analysed using content analysis. The number of positive words in each politician's statements were as follows:

| | | Political party | |
		Beige	Grey
Timing	Before	9	9
		11	9
		11	11
		7	12
		14	14
		12	11
		10	10
		8	7
		7	6
		8	9
	After	8	5
		6	6
		5	7
		8	4
		11	9
		9	10
		10	7
		8	5
		7	7
		7	6

Assume that each MP was randomly selected before the election and, if re-elected, was omitted from the random survey after the election (i.e. the data are independent).

(i) Carry out a two-way analysis of variance using ranks to determine whether there is an effect of party, timing (before or after the general election) or an interaction between the two.

	df	SS	MS	H	P
Party					
Timing					
Interaction					
Within					
Total					

(ii) Plot the rank means on an interaction line plot.

(iii) Using multiple comparison tests where appropriate, what can you conclude from these results?

7.6 The data below show the percentage of mammal, bird and reptile/amphibian species at risk (as a percentage of those species known) in eight Central American countries:

	Belize	Costa Rica	El Salvador	Guatemala	Honduras	Mexico	Nicaragua	Panama
Mammals	7.44	4.93	5.43	5.17	4.47	7.29	5.08	6.00
Birds	0.20	0.63	0.69	1.20	0.74	12.8	0.66	0.65
Reptiles/ Amphibians	6.02	2.44	5.38	3.30	4.13	3.90	4.07	3.27

(i) Carry out Friedman's matched group two-way analysis of variance using ranks to determine whether there is a difference between the percentage of the different groups of animals (mammals, birds and reptiles/amphibians) at risk (assume that we are not interested in differences between countries).

	Rank mean	Friedman's test		
Mammals		df:	F_r:	P:
Birds				
Reptiles/ amphibians				

(ii) Using multiple comparison tests where appropriate, what can you conclude from these results?

 **Glossary**

This section covers both the definitions of the statistical terms and the symbols used in this book

Definitions of statistical terms

absolute difference (Boxes 4.1 and 6.4). The positive value (i.e. ignoring the sign) of the difference between two numbers.

accidental sampling (Chapter 1). Individuals selected on the basis of first come, first sampled.

alternative hypothesis (Chapter 3). Prediction that samples differ, or variables are related or associated. Accepted if the **null hypothesis** is rejected. See also hypothesis, non-specific hypothesis, specific hypothesis.

analysis of variance (Chapters 5 and 7). A family of statistical tests for examining the differences in central tendency between samples. See also ANOVA, one-way analysis of variance, two-way analysis of variance, Kruskal–Wallis one-way analysis of variance using ranks, repeated measures ANOVA, Friedman's matched group analysis of variance using ranks.

ANOVA (Chapters 5 and 7). Acronym for **analysis of variance**, usually used for the parametric versions of the family of tests.

autocorrelation (Chapter 5). The situation where relationships exist between several variables, thereby making it difficult to separate out the effects of one variable on another.

bar chart (Chapter 2). Graph comprising bars which display the frequency (y axis) of one or more nominal variables (x axis) – see Figure 2.13. Alternatively, the frequencies (y axis) of a single discrete measured variable (x axis) may be displayed – see Figure 2.5. Bar charts may also be used to display the mean values of a measured variable (y axis) separated into one or more nominal categories (x axis). In the latter case, vertical lines representing standard errors or standard deviations may be added to the bars – see Figure 2.10a. Note that spaces separate the bars. Cf. histograms.

block (Figures 1.2, 1.3, and 7.4). Part of the design of an experiment or survey representing a division of space or time over which categories or treatments are allocated so as to take into account heterogeneity across space or time.

box and whisker plot (Figure 2.11). Graph comprising boxes which display the median value (centre line of box), upper and lower quartiles (top and bottom limits of box) and range (whiskers above and below the box) of an ordinal variable (*y* axis) separated into one or more nominal variables (*x* axis).

chi-square test (Chapter 6). Name often used for a statistical **test for association** (or independence) between two measured **frequency distributions**, the statistic from which (X^2 – often given the misleading designation of χ^2) approximates to the chi-square (χ^2) distribution. See also test for association/independence, goodness of fit test.

closed questions (Table 1.1). These offer respondents in a questionnaire a limited number of choices of response. Cf. open questions.

cluster sampling (Chapter 1). Sampling technique that makes use of situations where individual sampling units occur in groups. First groups (clusters) are randomly selected, then individuals are randomly selected within clusters.

coefficient of determination (Boxes 5.2 and 5.7). The proportion (r^2) or percentage (R^2) of variability in one variable that is shared by the variability in another.

confidence limits, also **confidence interval.** The range within which the mean lies with a given probability (e.g. 95% confidence limits of the mean contain 95% of possible sample means) – see Box 2.6. Confidence limits can also be used in regression to find, for a given probability, the range within which an estimated value of *x* lies – see Box 5.8. See also standard error.

confounding variable (Chapter 1). An independent variable, which is not under investigation, but which varies systematically with the independent variable under investigation.

content analysis (Chapter 1). Examination of text or interview transcript to identify characteristics (e.g. frequency, depth and position of responses).

contingency table. Table summarising two **frequency distributions**, one of which is laid out along the row axis and the other along the column axis, so that the cells of the table contain the frequencies held in common by the intersection of each row and column. Used in **cross-tabulation** displays (Chapter 1) and **testing for association/independence** of frequency data (e.g. chi-square test, Worked Example 6.1a).

continuous data (Chapter 1). Measurement data (**interval** or **ratio**) on a scale with an infinite number of points between any adjacent pair of values. Cf. discrete data.

control (Chapter 1). Part of an experiment set up to replicate all aspects of a treatment except that of interest to the investigator. Comparison of treatments against controls enables the determination of any significant treatment effects. Cf treatment.

correlation analysis (Chapter 5). Statistical test for relationships between two variables where cause and effect are not implied. See also Pearson's product moment correlation coefficient, Spearman's rank correlation coefficient. Cf. regression analysis.

critical level (Chapter 3). Arbitrary probability level, at values less than which an event is said to be significantly unlikely to occur (test statistics have associated probability distributions, and are usually considered unlikely to occur by chance if the probability is less than 0.05).

cross-tabulation. Table summarising two frequency distributions, one of which is laid out along the row axis and the other along the column axis, so that the cells of the table contain the frequencies held in common by the intersection of the row and column. The cells can also contain percentages based on either of the two distributions or on the total sample size – see Table 2.7. Mean or median values of a third variable could also comprise the cell contents – see Table 2.8. See also contingency table.

data. Information of any kind, often comprising scores, measurements or descriptions of individual sampling units.

degrees of freedom, df (Box 3.1). The number of data points in a sample that are free to vary. Calculated as the number of data points minus a value based on the number of parameters (e.g. the mean) that have been estimated from the sample.

denominator (Table A.1). The divisor in a fraction (i.e. the number below the line). Cf. numerator.

dependent variable (Chapters 1 and 5). One whose values are determined or hypothesised to be determined, at least in part, by the values of another variable. Cf independent variable.

derived data (Chapter 1). Data comprising proportions (or percentages) and rates which are either calculated from other measured data, or measured directly.

descriptive statistics (Chapter 2). Techniques to summarise and display data. Cf inferential statistics.

discrete or discontinuous data (Chapter 1). Data measured in units with distinct gaps between adjacent values (e.g. count, ordinal or nominal data). Cf. continuous data.

fixed variable (Chapter 1). Measurement, classification or score whose value is set as part of the experimental or survey design. See also variable. Cf. measured variable.

frequency distribution (Chapter 2). Summary of a data set (either graphically or in table form) which shows the number of occasions (frequency) that each value or range of values occurs.

Friedman's matched group analysis of variance using ranks (Chapter 7). **Nonparametric** statistical test to examine differences between more than two samples, while taking into account the existence of **matched** groups across the samples. Extension of the Wilcoxon matched-pairs test. See also repeated measures ANOVA. Cf. two-way analysis of variance, Kruskal–Wallis one-way analysis of variance using ranks.

goodness of fit test (Chapter 6). Statistical test for differences between a measured and a theoretical frequency distribution. See also chi-square test. Cf. test of association/independence.

histogram (Figure 2.1). Graph comprising bars which display the frequencies (y axis) of a measured (interval or ratio) variable (x axis). Note that there are no gaps between the bars on the x axis. See also frequency distribution. Cf. bar chart.

homogeneity (Chapter 6). Equality, often used in the context of between samples.

hypothesis (Chapter 3). Prediction about the differences between samples, or predictions about relationships or associations between variables. See also alternative hypothesis, null hypothesis.

independence of data (Chapter 1). Condition where the measured value (or score) of any individual does not affect the value (or score) of any other individual, either within or between samples. A condition for many statistical tests (except where matched or paired tests are being employed). See also matched data, unmatched data.

independent variable (Chapters 1 and 5). One whose values may determine the values of another variable, while being unaffected in return. Cf. dependent variable.

inferential statistics (Chapter 3). Techniques to identify differences between samples, or relationships or associations between variables. Cf. descriptive statistics.

interpolation (Worked Example 7.1d). Estimation of a statistic from a statistical table, where the exact degrees of freedom are not given, by calculating from the adjacent values. Note that exact probabilities can be interpolated in a similar manner.

interquartile range (Box 2.7). Describes the central portion of a sample containing 50% of the data points and lying between the 25% (lower **quartile**) and the 75% (upper **quartile**) points and surrounding (not necessarily symmetrically) the **median**. Usually used for skewed or ordinal data. Cf. standard deviation.

interval data (Box 1.3). Data on a measurement scale without an absolute zero point, so that data points can be distinguished from each other in terms of magnitude and directional differences, but ratios between points are not possible. Cf. nominal, ordinal, ratio data.

Kruskal–Wallis analysis of variance using ranks (Chapter 7). **Nonparametric** statistical test to examine differences between more than two samples comprising **unmatched** (independent) data. Cf. one-way analysis of variance.

Latin square (Figure 1.3). Stratified random arrangement of treatments in an experiment such that every treatment is represented once in every row and every column of the experimental design. Cf. randomised block design.

Line of best fit (Chapter 5). Calculated line shown on a **regression plot** which allows prediction of the dependent variable (displayed on the y axis) given knowledge of the independent variable (displayed on the x axis).

manipulative experiments (Chapter 1) Where one or more variables are manipulated to ascertain their effect on a measured variable while all other conditions are kept constant. Usually used to identify cause and effect. Cf. observational surveys.

Mann–Whitney U test (Chapter 4). **Nonparametric** statistical test for differences in central tendency between two samples comprising **unmatched** (independent) data. Cf. t test, Kruskal–Wallis analysis of variance using ranks, Wilcoxon matched-pairs test.

matched data (Chapters 4 and 7). The situation where each data point is not **independent** of all others, but pairs (or groups) of data points are linked by being taken from the same individual or sampling unit. See also independence of data.

mean, $\bar{x}$ (Chapter 2). The arithmetic average calculated as the sum of all the data points divided by the sample size. The most commonly used measure of central tendency for **interval** or **ratio** data. Cf. median, mode. See Chapter 2.

mean squares, MS (Chapter 7). A measure of the variation in a sample, calculated as the **sum of squares** divided by the degrees of freedom. Also called the **variance**, this term is most commonly employed in analysis of variance.

measured variable (Chapter 1). A measurement, classification or score recorded during an experiment or survey, the values of which are free to vary depending on the **fixed variables** set up in the experimental or survey design. See also variable.

median (Chapter 2). The middle value in a list of data ranked in order. The most commonly used measure of central tendency for **ordinal data**. Cf. mean, mode.

mode (Chapter 2). The most frequently occurring value. Often used for **nominal data**. Cf. mean, median.

model I regression analysis (Chapter 5). A form of regression analysis where the values of the independent variable are known without error or are fixed by the researcher. One example is **simple linear regression**. Cf. Model II regression analysis.

model II regression analysis (Chapter 5). A form of regression analysis where the values of the independent variable are not fixed by the researcher. Cf. Model I regression, simple linear regression.

nominal data (Box 1.3). Data on a categorical scale where values can be allocated to mutually exclusive categories and where categories have no magnitude or directional differences. Cf. ordinal, interval, ratio data.

nonparametric tests (Table 3.1). Statistical tests which do not require the data to be normally distributed, suitable for data measured at least on an **ordinal** scale (with the exception of tests of association/independence or goodness of fit, which use nominal data). Cf. parametric tests.

non-specific hypothesis (Chapter 3). A form of **alternative hypothesis** which predicts a difference between samples, or a relationship or association between variables without specifying the direction of such a difference, relationship or association. Cf. specific hypothesis.

normal distribution (Chapter 2). **Frequency distribution** with a symmetrical, bell-shaped frequency curve which has defined mathematical properties (e.g. that 68.27% of data points lie within one standard deviation either side of the mean). Cf. skewed distribution, Poisson distribution.

null hypothesis (Chapter 3). Prediction that samples do not differ, or that variables are not related or associated. This is the **hypothesis** at the start of a statistical test, and is rejected in favour of the **alternative hypothesis** if the probability is less than the critical value.

numerator (Table A.1). Number to be divided in a fraction (i.e. the number above the line). Cf. denominator.

observational surveys (Chapter 1). Surveys where variables are recorded without manipulation, usually in an attempt to examine real-life situations, but not appropriate for establishing cause and effect.

one-way analysis of variance (Chapter 7). Statistical test to examine differences between two or more samples where the data are **unmatched**. Although this is a generic term, it is usually taken to mean the **parametric** version (the nonparametric equivalent is **Kruskal–Wallis analysis of variance using ranks**). Cf. two-way analysis of variance.

open questions (Table 1.1). These offer respondents in a questionnaire survey a complete choice as to their response. Cf. closed questions.

ordinal data (Box 1.3). Data on a ranked scale where data points can be distinguished from each other in terms of directional differences, but not of magnitude. Cf. nominal, interval, ratio data.

paired t test (Chapter 4). Statistical test for the difference between two samples comprising **matched** pairs where the differences are **normally distributed**. Cf. Wilcoxon matched-pairs test, t test.

parameter. Measurement derived from a population (e.g. population mean or population standard deviation), values of which are usually estimated by taking a random sample of the population. Cf. statistic.

parametric tests (Table 3.1). Statistical tests which require the data to conform to the **normal distribution**. Cf. nonparametric tests.

Pearson's product moment correlation coefficient (Chapter 5). **Parametric** correlation analysis for relationships between two **normally distributed** variables. Cf. Spearman's rank correlation coefficient.

pie chart (Figure 2.13b). Circular graph which displays measurements or counts expressed as proportions of the whole, divided by categories of a **nominal** variable (slices of the pie).

point chart (Figure 2.10b). Graph comprising points (with or without vertical lines representing standard errors or standard deviations) which display the mean values of a variable (y axis) measured on an interval or ratio scale (or the rank mean values of a variable measured on an ordinal scale) separated by one or more nominal variables (x axis).

Poisson distribution (Chapters 3 and 6). **Frequency distribution** with defined mathematical characteristics (one of which is that the mean and variance are equal) which often pertains to counts of relatively rare items which are spatially and/or temporally distributed at random.

population (Chapter 1). Collection of all possible items (individual sample units) from which **samples** are taken.

power (Chapter 3). The ability of a test to find a difference between samples, or a relationship or association between variables, when one exists, i.e. to avoid **type II errors**.

predictive model (Chapter 5). The use of the values of one or more independent variables to predict the values of a dependent variable.

probability, P (Box 2.4). The chance of an event taking place, measured either as a proportion from 0 to 1 or as a percentage from 0 to 100% (where a value close to 0% is highly unlikely to occur by chance, while a value close to 100% is extremely likely to occur by chance). In statistical testing it is usually used to quantify the chance of the null hypothesis being true. Cf. critical value.

pseudoreplication (Chapter 1). Increasing the sample size by using non-independent data points. Should be avoided where possible.

quartile (Chapter 2). Position one-quarter (lower quartile, $Q_1 - 25\%$) or three quarters (upper quartile, $Q_3 - 75\%$) through a sample from the lowest value (usually used for data which are measured on an ordinal scale and/or are skewed). See also interquartile range.

quota sampling (Chapter 1). Sampling method where individuals are selected using a series of criteria such that once a predetermined number in any category has been reached, no further individuals are selected for that category.

random sampling (Figure 1.2). Selection of a sample where every member of the population has an equal chance of being chosen (may be with or without replacement, i.e. individuals may or may not be reselected). Often employs random numbers to aid selection. Cf. systematic sampling, stratified random sampling.

randomised block design (Figure 1.3). A form of block design where each position (e.g. row in a field) or time period (e.g. day) contains one of each of the treatments or categories in a random order. Cf. Latin square.

range (Chapter 2). The difference between the smallest and largest values of a sample. See also interquartile range.

rank mean (Chapters 4 and 7). The arithmetic average of a series of ranked values.

ranked values (Box 4.4). Each of the original values of a data set (measured at least on an ordinal scale) is replaced by a number indicating its position when the data set is ordered from low to high, taking into account ties. Many **nonparametric** statistical tests require data to be transformed in this way. See also ranking data.

ranking data. Methods of allocating ranks to data (required for many **nonparametric** tests). The procedure used varies depending on the test. For tests on unmatched samples (Mann–Whitney U test, Kruskal–Wallis, two-way analysis of variance using ranks), the whole data set is ranked as one. Where there are matched samples (Spearman's rank correlation coefficient, Friedman's matched groups analysis of variance using ranks), each variable (Spearman's) or group (Friedman's) is ranked separately. The Wilcoxon matched-pairs test is an exception, where the absolute differences between data pairs are ranked.

ratio data (Box 1.3). Data on a measurement scale with an absolute zero point so that data points can be distinguished from each other in terms of both magnitude and directional differences, and ratios between points are possible. Cf. nominal, interval, ordinal.

regression analysis (Chapter 5). Statistical tests for relationships where cause and effect are suspected between one normally distributed dependent variable and one independent variable (which is fixed by the researcher or known without error), or between two normally distributed variables. Also used when a predictive model is required. See also simple linear regression, model I and II regression analyses. Cf. correlation analysis.

regression plot (Figures 5.3 and 5.9). Graph combining a **scatterplot** and a **line of best fit** displaying the relationship between a normally distributed dependent variable and an independent variable (which is fixed by the researcher or known without error), or between two normally distributed variables (independent variable on the x axis and dependent variable on the y axis).

repeated measures analysis of variance (Chapter 7). Statistical test to examine differences between more than two samples while taking into account the existence of **matched** groups across the samples. The parametric version uses the same formulae as **two-way ANOVA** with single observations in each cell. The nonparametric version is **Friedman's matched group analysis of variance using ranks**.

residuals (Chapter 5). In **regression analysis**, the variation which remains when the regression variation has been taken into account (i.e. the difference between the *y* values and the regression line). Also sometimes used in ANOVA as a term for within-category variation.

sample (Chapter 1). A collection of individual items (that are assumed to be random and independent) drawn from a population, measurements from which are used to estimate the true measurements (**parameters**) of the population.

scatterplot (Figure 2.12). Graph with points displaying the relationship between two measured variables (independent variable on the *x* axis and dependent variable on the *y* axis).

semi-structured interviews (Chapter 1). Survey design which seeks responses from a number of defined questions while allowing the respondent to determine the nature of some of the information given. Cf. unstructured interviews.

simple linear regression (Chapter 5). A form of **regression analysis** where the values of the independent variable are known without error or are fixed by the researcher. A type of **model I regression analysis**. Cf. model II regression analysis.

skewed distribution (Chapter 2). An asymmetrical **frequency distribution** having a longer tail either to the left (negative skewness) or to the right (positive skewness). Cf. normal distribution.

snowball sampling (Chapter 1). A method of increasing the sample size by recruiting additional individuals from the initial (usually randomly) sampled respondents.

Spearman's rank correlation coefficient (Chapter 5). Nonparametric **correlation analysis** for examining relationships between two variables where one or both variables are non-normal or ordinal. Cf. Pearson's product moment correlation coefficient.

specific hypothesis (Chapter 3). A form of **alternative hypothesis** which predicts a difference between samples, or a relationship between variables, and specifies the direction of such a difference or relationship. Cf. non-specific hypothesis.

standard deviation (Box 2.3). Measure of the variation of a sample (interval or ratio) which provides a symmetrical range around the mean within which 68.27% of data points lie (if the data conform to a **normal distribution**). Calculated as the square root of the variance. Cf. interquartile range, standard error.

standard error (Box 2.5). Measure of the reliability of the estimation of a mean, comprising a symmetrical range around the mean (± one standard error) in which the probability of the true mean occurring is 68.27%. Cf. standard deviation, confidence limits.

statistic. Measurement derived from a sample (e.g. sample mean or sample standard deviation) intended to estimate the equivalent **parameter** of the population. Cf. test statistic.

statistical test (Chapter 3). Objective method of obtaining a probability that samples do not differ, variables are not related, or that frequency distributions are not associated. See also nonparametric tests, parametric tests.

stratified random sampling (Figure 1.2). Method of selecting individuals so that the full range of sampling possibilities are covered evenly by dividing the sample into blocks, within each of which samples are selected at random. Cf. random sampling, systematic sampling.

sum of squares, *SS* (Box 2.3). A measure of variation calculated as the deviations of each data point from the mean, squared and then summed. See also mean squares, variance.

systematic sampling (Figure 1.2). Selection of individuals such that the full range of possible individuals are covered evenly (e.g. by selecting every tenth individual). Cf. random sampling, stratified random sampling.

test for association/independence (Chapter 6). Statistical test for determining whether two or more **frequency distributions** of variables measured on a **nominal** scale are associated or independent. See also chi-square test. Cf. goodness of fit test.

test statistic (Chapter 3). Value calculated (or used) during a **statistical test**, either with a known distribution or one that approximates to a known distribution, from which the probability of the **null hypothesis** being true can be found.

transformation (Chapters 3 and 5). Mathematical manipulation of all of the data points in a sample to enable them to conform to the assumptions of a test (e.g. normal distribution or a linear relationship between variables).

treatment (Chapter 1). The manipulation of a sample in order to see whether the manipulation has an effect on a measured variable, ascertained by comparison with an unmanipulated sample. Cf. control.

t **test** (Chapter 4). Statistical test for differences between means in two **unmatched** samples comprising **normally distributed** data. Cf. *z* test, Mann–Whitney *U* test, analysis of variance, paired *t* test.

two-way analysis of variance (Chapter 7). Statistical test to simultaneously examine the effect of two independent variables and their interaction on a dependent variable. There are parametric and nonparametric versions, as well as versions appropriate when there are single observations in each cell. Cf. one-way analysis of variance.

type I error (Chapter 3). The probability of rejecting the **null hypothesis** when in fact it is true (i.e. producing falsely significant results). Cf. type II error.

type II error (Chapter 3). The probability of accepting the **null hypothesis** when in fact it is false (i.e. producing falsely insignificant results). Cf. type I error.

unmatched data (Chapters 4 and 7). The situation where data are independent of each other, having been sampled from different individuals or groups, so that the selection of any individual from one sample does not influence the selection of any other individual. See also independence of data. Cf. matched data.

unstructured interviews (Chapter 1). Survey design which seeks responses to a defined subject area without using fixed questions. Cf. semi-structured interviews.

variable (Chapter 1). Characteristic (comprising measurements, classifications or scores) that varies between individuals within a population. See also fixed variable, measured variable, dependent variable, independent variable.

variance (Box 2.3). Measure of variation in a sample, standardised by the size of the sample. Calculated as the **sum of squares** divided by the degrees of freedom. See also standard deviation, mean squares.

Wilcoxon matched-pairs test (Chapter 4). **Nonparametric** statistical test for differences between two samples comprising data in **matched pairs**. Cf. Mann–Whitney U test, Friedman's matched group analysis of variance using ranks, paired t test.

z **test** (Chapter 4). Statistical test for differences between the means of two **unmatched, normally distributed** samples where the size of each sample is large (i.e. exceeds 30). A large-sample version of the t test.

Mathematical symbols

a	Intercept of a line on a graph (see Box 5.4)
b	Gradient of a line on a graph (see Box 5.4)
c	Number of columns in a contingency table (see Box 6.2) or a two-way analysis of variance (see Figure 7.4)
CI	Confidence intervals (see confidence limits)
CL	Confidence limits (see Box 2.6)
d	The difference between pairs of numbers in a paired t-test calculation (see Box 4.5)
df	Degrees of freedom (see Box 3.1)
E	Expected value (see Box 6.1)
f	Frequency as used in frequency tables or histograms (see Table 2.1)

F Test statistic for the F test used in parametric analysis of variance (see Tables 5.4 and 7.2)

F_r Test statistic for Friedman's matched group analysis of variance using ranks (see Box 7.9)

g Number of groups used in a matched test (see Box 7.9)

G Test statistic for G test for associations between frequency distributions and goodness-of-fit tests (see Chapter 6)

H Test statistic for the Kruskal–Wallis one-way analysis of variance using ranks, and two-way analysis of variance using ranks (see Box 7.4)

H_0 Null hypothesis (Chapter 3)

k Number of samples (see Chapter 7, e.g. Box 7.1)

MS Mean square (see Table 7.2)

MSD Minimum significant difference (e.g. see Box 7.2)

n number of individuals in the sample (see Chapter 2, e.g. Worked Example 2.1)

O Observed value (see Box 6.2)

P Probability (see Boxes 2.4 and 3.2)

q Test statistic used in the Tukey, Tukey–Kramer and Nemenyi tests for minimum significant differences (e.g. see Box 7.2)

Q Test statistic used in Dunn's test for minimum significant differences (see Box 7.6)

Q_1 Lower quartile (see Box 2.7)

Q_3 Upper quartile (see Box 2.7)

r Correlation coefficient (see Box 5.1); also used to indicate the number of rows in a contingency table (see Box 6.2) or a two-way analysis of variance (see Figure 7.4)

r_S Spearman's rank correlation coefficient (see Box 5.3)

r^2 Coefficient of determination expressed as a proportion (see Box 5.2)

R^2 Coefficient of determination expressed as a percentage (see Box 5.2)

s Standard deviation (see Box 2.3)

s^2 Variance (see Box 2.3)

SE Standard error (see Box 2.5)

SS Sum of squares (see Boxes 2.3 and 7.1)

t t statistic (see Boxes 4.1 and 4.5)

T Test statistic for Wilcoxon matched pairs test (see Worked Example 4.4)

U Test statistic for Mann–Whitney U test (see Box 4.3)

x Data point (see Table 2.1)

x' The value of x for which we wish to estimate y when predicting from a regression line (see Box 5.8)

$\bar{x}$ Sample mean (arithmetic average of x: see Box 2.1)

X^2 Test statistic for the chi-square test (see Box 6.2)

(x, y) Coordinate of a point on a graph with x and y axes (Box 5.4)

$\bar{y}$ Mean of y (see Chapter 5)

z Test statistic for the z test (see Box 4.2)

μ Population mean (see Chapter 2)

σ_n Population standard deviation (sometimes used as an alternative to s: see Box 2.3)

σ_{n-1} Sample standard deviation (sometimes used as an alternative to s: see Box 2.3)

χ^2 Chi-square (see Box 6.2)

$\sum R$ Sum of ranks (see Box 4.3)

∞ Infinity

Appendix A

Basic mathematical methods and notation

Table A.1 lists the basic mathematical methods and notation which you may need to remind yourself of or familiarise yourself with before attempting to calculate by hand the formulae given in this book.

Table A.1	Basic mathematical methods and notation	
Type of calculation	*Notation*	*Examples*
Addition	$a + b = c$	$4 + 3 = 7$
Subtraction	$a - b = c$	$4 - 3 = 1$
Multiplication	$a \times b = c$ or $ab = c$ (also called the product of a and b)	$4 \times 3 = 12$
Division	$a \div b = c$ or $\dfrac{a}{b} = c$ or $a / b = c$ (where a is the **numerator** and b is the **denominator**)	$10 \div 2 = 5$ or $\dfrac{10}{2} = 5$ or $10 / 2 = 5$
Using brackets	$(a + b) \times (c - d) = e$ or $(a + b)(c - d) = e$	$(2 + 3) \times (6 - 4) = 10$ Calculations within the brackets must be completed first i.e. $(2 + 3) \times (6 - 4) = 5 \times 2 = 10$
The square of a number	$a^2 = a \times a = b$	$4^2 = 4 \times 4 = 16$ Where a squared value is part of a longer formula, it should be treated as if it is in brackets, i.e. calculate the square value first before continuing with the formula, e.g. $4^2 + 6 = (4 \times 4) + 6$ $= 16 + 6 = 22$

The square root of a number	$\sqrt{a} = b$	$\sqrt{4} = 2$ i.e. find the number which when multiplied by itself equals the number under the square root sign (here, $2 \times 2 = 4$)
Powers of 10 and their use by calculators and computers	10^a or 10^{-a}	$10^3 = 1000$ or $10^{-3} = 0.001$ Some calculators and computers report small numbers in this form, e.g. 2.3 E-03 $= 2.3 \times 10^{-3}$ $= 2.3 \times 0.001 = 0.0023$ Similarly: $2.3 \times 10^3 =$ $2.3 \times 1000 = 2300$
Calculating the sum of a list of numbers	$\sum x$ Sum of x, where x represents a list of numbers, e.g. 1, 2, 3, 4	$1 + 2 + 3 + 4 = 10$
Taking the absolute value (i.e. ignoring the sign)	$\lvert a - b \rvert = c$	$\lvert 2 - 3 \rvert = 1$ $2 - 3 = -1$, but taking the absolute value means ignoring the minus sign
The 'greater than' symbol	$a > b$ The symbol $\geq$ represents greater than or equal to	$4 > 2$ i.e. 4 is greater than 2
The 'less than' symbol	$a < b$ The symbol $\leq$ represents less than or equal to	$3 < 5$ i.e. 3 is less than 5

Appendix B
Using statistical software

Most users of this book will employ statistical programs on computers to analyse their data. Not only must you be able to select the correct test (see Chapter 3) and then interpret the output (see Chapters 4–7), but it is essential that you enter the data in the way required by the program for each statistical test. The commonest computer programs in use these days work in similar ways and, once you master one, getting to grips with another is reasonably straightforward. In all cases you should ensure that you understand the way that the program uses your data and the distinctions between the various forms of output (e.g. producing test statistics and associated probabilities, with and without correction factors for small sample sizes or tied ranks).

Most software uses a similar data entry format to the data recording sheet shown in Figure 1.6. The ways in which such data are used by different tests are shown in Tables B.1 (*t* test and Mann–Whitney *U* test), B.2 (paired *t* test and Wilcoxon matched-pairs test), B.3 (correlations and regression), B.4 and B.5 (tests for association between frequencies), B.6 and B.7 (analyses of variance). It is good practice to have an additional column which is not used within the analysis but which records the individual number (see Figure 1.6) to assist in data checking and cross-referencing against the raw data in laboratory or field recording sheets.

In general terms, the fixed variables will be those determined as part of the experimental or survey design, while the measured variables will be those recorded as part of the research. The fixed variables are usually used to separate data (examining one group of data separately from another), to provide groups for comparison (e.g. in difference testing or tests of association) and as independent variables for examining relationships between variables. The measured variables will tend to be those used within comparisons of differences, relationships and tests of association. However, measured variables which are collected on a nominal scale may, in addition to use in tests of association, be used to separate the data into groups (either for separate testing of different groups or to provide groups for comparisons). Letter codes or names are more convenient for labelling nominal variables (e.g. 'F' for female and 'M' for male), rather than using numeric values (e.g. where the code '1' indicates female and '2' indicates male), because it is easier to interpret the output (i.e. you do not have to remember what each code stood for). This also avoids the temptation to use numeric codes as if they were real measurements in an analysis. Unfortunately, some programs require nominal variables to be labelled with numbers. It is important to check the requirements of the program you are using before entering the data.

Note that it is up to the user to decide which statistical test is most appropriate (i.e. whether the data are measured on an appropriate scale and are normally distributed or not). Computers will merely do as instructed, irrespective of whether it makes any sort of statistical sense.

Table B.1 *Data entry for unmatched comparisons between two samples (t test and Mann–Whitney U test)*

Fixed variable 1	Fixed variable 2	Measured variable 1	Measured variable 2	etc.
A	.	6.4	.	
A	.	7.2	.	
B	.	5.2	.	
B	.	8.1	.	
B	.	5.8	.	
A	.	4.6	.	
A	.	3.9	.	
etc.				

In Table B.1, fixed variable columns contain the categories to be compared (A vs B) while the measured variable columns contain the values which will comprise the comparisons. Each row represents an individual.

So for *t* tests, a program would compare the means of measured variable 1 for each category (A and B) of fixed variable 1, and calculate the test statistic (*t*) and its associated probability.

Similarly, for Mann–Whitney *U* tests, the program would compare the ranked values of measured variable 1 for each category (A and B) of fixed variable 1, and calculate the test statistic (*U* and/or *z*) and its associated probability.

Table B.2 *Data entry for matched comparisons between two samples (paired t test and Wilcoxon matched-pairs test)*

Fixed variable 1	Fixed variable 2	Measured variable 1	Measured variable 2	etc.
A	.	6.4	3.6	
A	.	7.2	5.6	
B	.	5.2	6.2	
B	.	8.1	7.4	
B	.	5.8	4.6	
A	.	4.6	4.8	
A	.	3.9	2.9	
etc.				

In Table B.2, data are entered in columns, within which a fixed variable column may contain the codes for the categories with which the data may be separated if required (e.g. fixed variable 1 could be used to separate the data into category A and category B). The two measured variable columns contain the pairs of values which will comprise the comparisons. Each row represents an individual or matched group.

So for paired *t* tests, the program would generate the mean of the difference between matched pairs from measured variable 1 and measured variable 2 and calculate the test statistic (*t*) and its associated probability.

Similarly, for Wilcoxon matched-pairs tests, the program would rank the differences between matched pairs of measured variable 1 and measured variable 2 and calculate the test statistic (*T* and/or *z*) and its associated probability.

Table B.3 *Data entry for relationships between two variables (Pearson's product moment correlation coefficient, Spearman's rank correlation coefficient and regression analysis)*

Fixed variable 1	Fixed variable 2	Measured variable 1	Measured variable 2	etc.
A	5	6.4	3.6	
A	10	7.2	5.6	
B	15	5.2	6.2	
B	20	8.1	7.4	
B	25	5.8	4.6	
A	30	4.6	4.8	
A	35	3.9	2.9	
etc.				

In Table B.3, fixed variable columns contain the categories with which the data may be separated (e.g. fixed variable 1 could be used to separate the data into category A and category B – note that it is often essential for different categories to be separated before relationship testing) and independent variables for regression analysis (e.g. fixed variable 2). Each row represents an individual.

For correlations, a program would use paired values of measured variable 1 and measured variable 2 to calculate the test statistic (r) and its associated probability. Pearson's product moment correlation coefficient uses the actual data values, whereas Spearman's rank correlation coefficient calculates the test statistic (r_S) on the ranks. Although, fixed variables are not free to vary and, strictly speaking, correlation examines how variables covary, fixed variables may also be used in correlations.

For regression analysis, fixed variable 1 would be used as the independent variable, with one of the measured variables as the dependent variable. The program would calculate the test statistic (F or t) and its associated probability, and generate an equation describing the relationship.

Table B.4 *Data entry for frequency analysis: coded raw data (test for associations between frequency distributions)*

Fixed variable 1	Fixed variable 2	Measured variable 1	Measured variable 2	etc.
A	•	W	P	
A	•	T	P	
B	•	T	Q	
B	•	U	R	
B	•	W	R	
A	•	W	Q	
A	•	T	R	
etc.				

In Table B.4, both the fixed and measured variable columns may contain nominal data to be compared. Each row represents an individual.

The program first calculates the frequencies of occurrence of each combination of codes from the selected two variables (unlike inputting the data directly in a contingency table as in Table B.5). After producing a contingency table, the program then calculates the test statistic (X^2) and its probability. So if the frequencies of fixed variable 1 (A or B) and measured variable 1 (T, U, or W) were examined, the results would be a 2 × 3 contingency table containing the frequencies of AT, AU, AW, BT, BU and BW, from which the program would calculate the expected values, test statistic (X^2) and its associated probability. If the frequencies of measured variable 1 and measured variable 2 were examined, the resulting 3 × 3 contingency table would compare the observed to the expected frequencies of TP, TQ, TR, UP, UQ, UR, WP, WQ and WR.

Table B.5 *Data entry for frequency analysis: contingency table (test for association between frequency distributions and goodness of fit)*

Variable 1	Variable 2	Variable 3	Variable 4	etc.
23	30			
65	50			
12	20			

In Table B.5, data are entered in the form of a contingency table (see Worked Example 6.1a) in which the columns contain the contingency column data and the rows contain the contingency row data. It is not usually possible to name the rows in a computer data sheet (you will have to remember which is which).

Unlike coded raw data (Table B.4), the computer program is given the frequencies already calculated and simply uses these to calculate the expected frequencies, the test statistic (X^2) and its associated probability.

For a goodness of fit test, one column would contain the observed values, and the second would contain the expected values from the theoretical distribution.

Table B.6 *Data entry for unmatched comparisons between more than two samples (one-way ANOVA, two-way ANOVA, Kruskal–Wallis one-way analysis of variance using ranks and two-way ANOVA using ranks)*

Fixed variable 1	Fixed variable 2	Measured variable 1	Transformed variable 1	etc.
A	M	6.4	9	
A	M	7.2	10	
A	N	5.2	6	
A	N	8.1	11	
B	M	5.8	7	
B	M	4.6	4	
B	N	3.9	2.5	
B	N	6.3	8	
C	M	4.7	5	
C	M	2.8	1	
C	N	8.4	12	
C	N	3.9	2.5	
etc.				

In Table B.6, fixed variable columns contain the codes for the categories to be compared (e.g. A vs B vs C of fixed variable 1, and M vs N of fixed variable 2) while the measured variable columns contain the values which will comprise the comparisons. Each row represents an individual.

For one-way ANOVA, a program would compare the values of measured variable 1 for each category (A, B and C) of fixed variable 1 and calculate the sums of squares, the test statistic (F) and its associated probability. For Kruskal–Wallis one-way analysis of variance using ranks, the program would rank the data and provide the test statistic (H) and its associated probability.

For two-way ANOVA, a program would compare the values of measured variable 1 for each category of both fixed variable 1 (A, B and C) and fixed variable 2 (M and N), and calculate the test statistics (F) and their associated probabilities. The commonly used programs do not include a two-way analysis of variance using ranks. However, you can instruct the computer to rank the data into a new column (e.g. transformed variable 1), perform a two-way ANOVA, and then calculate the H statistic by hand from the sums of squares using the formula in Box 7.8.

Table B.7 *Data entry for comparisons between more than two matched groups (repeated measures ANOVA and Friedman's matched group analysis of variance using ranks)*

Fixed variable 1	Measured variable 1	Measured variable 2	Measured variable 3	etc.
•	5.3	6.4	3.6	
•	8.9	7.2	5.6	
•	7.6	5.2	6.2	
•	8.4	8.1	7.4	
•	3.7	5.8	4.6	
•	5.1	4.6	4.8	
•	6.8	3.9	2.9	
etc.				

In Table B.7, data are entered in columns, within which fixed variable columns can contain the codes for each group of data, and the categories with which the data may be separated if required. The measured variable columns contain the values which will comprise the comparisons. Each row represents an individual or a matched group.

For repeated measures ANOVA, the program would compare the measured variables and calculate the sums of squares, the test statistic (F) and its associated probability.

For Friedman's matched group analysis of variance using ranks, the program first ranks the differences within matched groups (i.e. rows) across measured variable 1, measured variable 2 and measured variable 3, and then calculates the test statistic (F_r) and its associated probability. Note that some programs calculate the test statistic as χ^2.

Appendix C
Further reading

This book is intended as an introduction to a wide range of statistical tests that are required for environmental investigations. There are many other tests which you may come across in reading the literature and in the context of more specialist use. It would be impossible to do other than give a flavour of where to go to gain help in areas beyond the scope of this book. The bibliography below is highly selective and contains those texts we, our colleagues and students have found useful in the past.

Selected texts

Barnard C., Gilbert F. and McGregor P. (1993) *Asking Questions in Biology*. Longman, Edinburgh.
A useful introduction to experimental and survey design, especially useful for a discussion of specific hypotheses for nonparametric tests.

de Vaus D. A. (1996) *Surveys in Social Science* (4th edition), Social Research Today 5. UCL Press, London.
Good coverage of many elements of constructing and administering questionnaires.

Haining R. (1993) *Spatial Data Analysis in the Social and Environmental Sciences*. Cambridge University Press, Cambridge.
Covers most of the tests needed for spatial statistics, using relevant examples.

Lindsay J. M. (1997) *Techniques in Human Geography*. Routledge, London.
Good coverage of project design, questionnaires and interviews.

Manly B. J. F. (1992) *The Design and Analysis of Research Studies*. Cambridge University Press, Cambridge.
Very good coverage of experimental and survey design.

Manley B. F. J. (1994) *Multivariate statistical methods* (2nd edition). Chapman & Hall, London.
Covers the major areas of multivariate statistics in an accessible style.

Neave H. R. and Worthington P. L. (1988) *Distribution-free Tests*. Unwin Hyman, London.
Reference text covering the main nonparametric tests in an accessible style.

Robinson G. M. (1998) *Methods and Techniques in Human Geography*. Wiley, Chichester.
Covers quantitative (including spatial and multivariate methods) and qualitative techniques.

Siegel S. and Castellan N. J. (1988) *Nonparametric statistics for the behavioral sciences* (2nd edition). McGraw-Hill, New York.
A good reference with a very thorough coverage of nonparametric tests.

Sokal R. R. and Rohlf F. J. (1995) Biometry, (3rd edition). W. H. Freeman & Co., New York.
Technical approach but a good reference text for the initiated, with good coverage of parametric tests and some nonparametric tests.

Tabachnick B. G. and Fidell L. S. (1996) *Using Multivariate Statistics* (3rd edition). HarperCollins, New York.
Very clear account of the major multivariate statistics, followed by more technical explanations. Also gives some examples of the output from some commonly used statistical packages.

Watts S. and Halliwell L. (eds) (1996) *Essential Environmental Science*. Routledge, London.
Good introduction to relevant experimental and survey methods.

Zar J. H. (1999) *Biostatistical Analysis* (4th edition). Prentice Hall Inc., Upper Saddle River, NJ.
Good coverage of many parametric and nonparametric tests. Technical approach, but slightly more accessible, although (in places) very slightly less complete than Sokal and Rohlf (1995).

Statistical tables

Many of the above books include statistical tables. Sokal and Rohlf (1995) is an exception, and uses a separate book of tables (see below). Separate tables may be required for some data sets or tests which are not covered by those in the back of the above books. The following are three examples of books containing only statistical tables:

Neave H. R. (1995a) *Elementary Statistics Tables*. Routledge, London.
Well-laid-out tables which are reasonably comprehensive.

Neave H. R. (1995b) *Statistics Tables*. Routledge, London.
Similar to Neave (1995a): slightly more comprehensive, but less well laid out.

Rohlf F. J. and Sokal R. R. (1995) *Statistical Tables* (3rd edition). W. H. Freeman & Co., New York.
A good companion to Sokal and Rohlf (1995) containing a comprehensive selection of tables.

Appendix D
Statistical tables

This appendix provides the statistical tables that support the tests described in this book. Each provides a range of values that should be sufficient for most student practicals and projects. For larger data sets it is unlikely that the analysis would be carried out by hand, and an appropriate computer program should give exact probabilities. Any probabilities for degrees of freedom not covered by these tables may be interpolated (see Worked Example 7.1d) so long as degrees of freedom above and below the required value are present. In the event of needing more complete tables, consult those suggested in Appendix C. The tables included in this appendix give P values for non-specific tests: more comprehensive tables may be required if specific hypotheses are being tested. Throughout these tables, the test statistic for $P = 0.05$ is given in bold. Where appropriate, the title of the table gives a reminder of the calculations of the degrees of freedom (df), where n indicates the number of data points, k the number of samples, r the number of rows, c the number of columns and g the number of groups.

Table D.1 *Values of z. These values are used to calculate the 95% confidence limits of the mean when the sample size (n) is large (>30). They are also used in a z test (see Box 4.2). When used in z tests, the results are significant if the calculated value of z is higher than the table value*

$P = 0.1$	$P = 0.05$	$P = 0.01$	$P = 0.001$
1.645	**1.960**	2.576	3.291

Table D.2 *Values of t for use in calculating confidence limits and in unpaired and paired t tests. If using these values to calculate 95% confidence limits of the mean (see Box 2.6), then the df are n – 1. If this table is being used for an unpaired t test (see Box 4.1), the df are $n_1 + n_2 – 2$ (where n_1 and n_2 are the sizes of samples 1 and 2, respectively). If this table is being used for a paired t test (Box 4.5), the df are n – 1 (where n is the number of data pairs). When used in t tests (paired or unpaired), the results are significant if the calculated value of t is higher than the table value*

df	t		df	t	
	P = 0.05	P = 0.01		P = 0.05	P = 0.01
1	12.706	63.657	19	2.093	2.861
2	4.303	9.925	20	2.086	2.845
3	3.182	5.841	21	2.080	2.831
4	2.776	4.604	22	2.074	2.819
5	2.571	4.032	23	2.069	2.807
6	2.447	3.707	24	2.064	2.797
7	2.365	3.499	25	2.060	2.787
8	2.306	3.355	26	2.056	2.779
9	2.262	3.250	27	2.052	2.771
10	2.228	3.169	28	2.048	2.763
11	2.201	3.106	29	2.045	2.756
12	2.179	3.055	30	2.042	2.750
13	2.160	3.012	40	2.021	2.704
14	2.145	2.977	50	2.009	2.678
15	2.131	2.947	60	2.000	2.660
16	2.120	2.921	100	1.984	2.626
17	2.110	2.898	120	1.980	2.617
18	2.101	2.878	∞	1.960	2.576

Table D.3 *Mann–Whitney U values. The upper table values (in bold) are for P = 0.05, while the lower values are for P = 0.01. The results are significant if the lowest calculated U value is equal to or lower than the table value*

n_L (number of items in the larger sample) — columns. n_S (number of items in the smaller sample) — rows. For each n_S the first line (bold) is P = 0.05 and the second line is P = 0.01.

n_S	P	2	3	4	5	6	7	8	9	10	11	12	13	14	15	16	17	18	19	20	21	22	23	24	25
2	0.05	–	–	–	–	–	–	**0**	**0**	**0**	**0**	**1**	**1**	**1**	**1**	**1**	**2**	**2**	**2**	**2**	**3**	**3**	**3**	**3**	**3**
	0.01	–	–	–	–	–	–	–	–	–	–	–	–	–	–	–	–	–	0	0	0	0	0	0	0
3	0.05	–	–	–	**0**	**1**	**1**	**2**	**2**	**3**	**3**	**4**	**4**	**5**	**5**	**6**	**6**	**7**	**7**	**8**	**8**	**9**	**9**	**10**	**10**
	0.01	–	–	–	–	–	–	–	0	0	0	1	1	1	2	2	2	3	3	3	4	4	4	5	5
4	0.05	–	–	**0**	**1**	**2**	**3**	**4**	**4**	**5**	**6**	**7**	**8**	**9**	**10**	**11**	**11**	**12**	**13**	**14**	**15**	**16**	**17**	**17**	**18**
	0.01	–	–	–	–	0	0	1	1	2	2	3	3	4	5	5	6	6	7	8	8	9	9	10	10
5	0.05				**2**	**3**	**5**	**6**	**7**	**8**	**9**	**11**	**12**	**13**	**14**	**15**	**17**	**18**	**19**	**20**	**22**	**23**	**24**	**25**	**27**
	0.01				0	1	1	2	3	4	5	6	7	7	8	9	10	11	12	13	14	14	15	16	17
6	0.05					**5**	**6**	**8**	**10**	**11**	**13**	**14**	**16**	**17**	**19**	**21**	**22**	**24**	**25**	**27**	**29**	**30**	**32**	**33**	**35**
	0.01					2	3	4	5	6	7	9	10	11	12	13	15	16	17	18	19	21	22	23	24
7	0.05						**8**	**10**	**12**	**14**	**16**	**18**	**20**	**22**	**24**	**26**	**28**	**30**	**32**	**34**	**36**	**38**	**40**	**42**	**44**
	0.01						4	6	7	9	10	12	13	15	16	18	19	21	22	24	25	27	29	30	32
8	0.05							**13**	**15**	**17**	**19**	**22**	**24**	**26**	**29**	**31**	**34**	**36**	**38**	**41**	**43**	**45**	**48**	**50**	**53**
	0.01							7	9	11	13	15	17	18	20	22	24	26	28	30	32	34	35	37	39
9	0.05								**17**	**20**	**23**	**26**	**28**	**31**	**34**	**37**	**39**	**42**	**45**	**48**	**50**	**53**	**56**	**59**	**62**
	0.01								11	13	16	18	20	22	24	27	29	31	33	36	38	40	43	45	47
10	0.05									**23**	**26**	**29**	**33**	**36**	**39**	**42**	**45**	**48**	**52**	**55**	**58**	**61**	**64**	**67**	**71**
	0.01									16	18	21	24	26	29	31	34	37	39	42	44	47	50	52	55
11	0.05										**30**	**33**	**37**	**40**	**44**	**47**	**51**	**55**	**58**	**62**	**65**	**69**	**73**	**76**	**80**
	0.01										21	24	27	30	33	36	39	42	45	48	51	54	57	60	63
12	0.05											**37**	**41**	**45**	**49**	**53**	**57**	**61**	**65**	**69**	**73**	**77**	**81**	**85**	**89**
	0.01											27	31	34	37	41	44	47	51	54	58	61	64	68	71
13	0.05												**45**	**50**	**54**	**59**	**63**	**67**	**72**	**76**	**80**	**85**	**89**	**94**	**98**
	0.01												34	38	42	45	49	53	57	60	64	68	72	75	79
14	0.05													**55**	**59**	**64**	**69**	**74**	**78**	**83**	**88**	**93**	**98**	**102**	**107**
	0.01													42	46	50	54	58	63	67	71	75	79	83	87
15	0.05														**64**	**70**	**75**	**80**	**85**	**90**	**96**	**101**	**106**	**111**	**117**
	0.01														51	55	60	64	69	73	78	82	87	91	96
16	0.05															**75**	**81**	**86**	**92**	**98**	**103**	**109**	**115**	**120**	**126**
	0.01															60	65	70	74	79	84	89	94	99	104
17	0.05																**87**	**93**	**99**	**105**	**111**	**117**	**123**	**129**	**135**
	0.01																70	75	81	86	91	96	102	107	112
18	0.05																	**99**	**106**	**112**	**119**	**125**	**132**	**138**	**145**
	0.01																	81	87	92	98	104	109	115	121
19	0.05																		**113**	**119**	**126**	**133**	**140**	**147**	**154**
	0.01																		93	99	105	111	117	123	129
20	0.05																			**127**	**134**	**141**	**149**	**156**	**163**
	0.01																			105	112	118	125	131	138
21	0.05																				**142**	**150**	**157**	**165**	**173**
	0.01																				118	125	132	139	146
22	0.05																					**158**	**166**	**174**	**182**
	0.01																					133	140	147	155
23	0.05																						**175**	**183**	**192**
	0.01																						148	155	163
24	0.05																							**192**	**201**
	0.01																							164	172
25	0.05																								**211**
	0.01																								180

Table D.4 *Values of* T *for the Wilcoxon matched-pairs test. The results are significant if the calculated value of* T *is less than or equal to the table value*

n (number of items minus any zero differences)	T P = 0.05	P = 0.01	n (number of items minus any zero differences)	T P = 0.05	P = 0.01
6	0	–	16	29	19
7	2	–	17	34	23
8	3	0	18	40	27
9	5	1	19	46	32
10	8	3	20	52	37
11	10	5	21	58	42
12	13	7	22	65	48
13	17	9	23	73	54
14	21	12	24	81	61
15	25	15	25	89	68

Table D.5 *Values of Pearson's product moment correlation coefficient* (r). *The results are significant if the calculated value of* r *is higher than the table value*

df (n – 2)	r P = 0.05	P = 0.01	df (n – 2)	r P = 0.05	P = 0.01
1	0.997	1.000	16	0.468	0.590
2	0.950	0.990	17	0.456	0.575
3	0.878	0.959	18	0.444	0.561
4	0.811	0.917	19	0.433	0.549
5	0.754	0.874	20	0.423	0.537
6	0.707	0.834	21	0.413	0.526
7	0.666	0.798	22	0.404	0.515
8	0.632	0.765	23	0.396	0.505
9	0.602	0.735	24	0.388	0.496
10	0.576	0.708	25	0.381	0.487
11	0.553	0.684	26	0.374	0.479
12	0.532	0.661	27	0.367	0.471
13	0.514	0.641	28	0.361	0.463
14	0.497	0.623	29	0.355	0.456
15	0.482	0.606	30	0.349	0.449

Table D.6 *Selected values of Spearman's rank correlation coefficient (r_S). The results are significant if the calculated value of r_S is higher than the table value*

n	r_S		n	r_S	
	$P = 0.05$	$P = 0.01$		$P = 0.05$	$P = 0.01$
5	**1.000**	–	18	**0.472**	0.600
6	**0.886**	1.000	19	**0.460**	0.584
7	**0.786**	0.929	20	**0.447**	0.570
8	**0.738**	0.881	21	**0.436**	0.556
9	**0.700**	0.833	22	**0.425**	0.544
10	**0.648**	0.794	23	**0.416**	0.532
11	**0.618**	0.755	24	**0.407**	0.521
12	**0.587**	0.727	25	**0.398**	0.511
13	**0.560**	0.703	26	**0.390**	0.501
14	**0.538**	0.679	27	**0.383**	0.492
15	**0.521**	0.654	28	**0.375**	0.483
16	**0.503**	0.635	29	**0.368**	0.475
17	**0.488**	0.618	30	**0.362**	0.467

Table D.7 ***Values of F for use in regression analysis, one-way ANOVA and two-way ANOVA. The upper table values (in bold) are for* P = *0.05, while the lower values are for* P = *0.01. Look up the table values using the df for the explained and unexplained variation. For regression analysis (see Table 5.4), the df for the explained variation are the regression df (1) and those for the unexplained variation are the residual df* (n – *2). For one way ANOVA (see Table 7.2), the df for the explained variation are the between df (*k – 1*), while those for the unexplained variation are the within df (*[n_T – 1] – [k – 1]*). For two-way ANOVA (see Table 7.9), the df for the explained variation are the df associated with the variable (*r – 1 *or* c – 1*) or the interaction between the variables (*[r – 1] [c – 1]*), while those for the unexplained variation are the within df (*r × c [n – 1]*). For two-way ANOVA with single observations in each cell (see bottom of Worked Example 7.7), the df for the explained variation are the df associated with the variable (*r – 1 *or* c – 1*), while those for the unexplained variation are the remainder df (*[r – 1][c – 1]*). The results are significant if the calculated value of* F *is higher than the table value***

df for the unexplained variation (e.g. residual df or within df)	df for the explained variation (e.g. regression df or between df): 1	2	3	4	5
1	**161**	**200**	**216**	**225**	**230**
	4052	5000	5403	5625	5764
2	**18.5**	**19.0**	**19.2**	**19.2**	**19.3**
	98.5	99.0	99.2	99.2	99.3
3	**10.1**	**9.55**	**9.28**	**9.12**	**9.01**
	34.1	30.8	29.5	28.7	28.2
4	**7.71**	**6.94**	**6.59**	**6.39**	**6.26**
	21.2	18.0	16.7	16.0	15.5
5	**6.61**	**5.79**	**5.41**	**5.19**	**5.05**
	16.3	13.3	12.1	11.4	11.0
6	**5.99**	**5.14**	**4.76**	**4.53**	**4.39**
	13.7	10.9	9.78	9.15	8.75
7	**5.59**	**4.74**	**4.35**	**4.12**	**3.97**
	12.2	9.55	8.45	7.85	7.46
8	**5.32**	**4.46**	**4.07**	**3.84**	**3.69**
	11.3	8.65	7.59	7.01	6.63
9	**5.12**	**4.26**	**3.86**	**3.63**	**3.48**
	10.6	8.02	6.99	6.42	6.06
10	**4.96**	**4.10**	**3.71**	**3.48**	**3.33**
	10.0	7.56	6.55	5.99	5.64
11	**4.84**	**3.98**	**3.59**	**3.36**	**3.20**
	9.65	7.21	6.22	5.67	5.32
12	**4.75**	**3.89**	**3.49**	**3.26**	**3.11**
	9.33	6.93	5.95	5.41	5.06
13	**4.67**	**3.81**	**3.41**	**3.18**	**3.03**
	9.07	6.70	5.74	5.21	4.86
14	**4.60**	**3.74**	**3.34**	**3.11**	**2.96**
	8.86	6.51	5.56	5.04	4.69
15	**4.54**	**3.68**	**3.29**	**3.06**	**2.90**
	8.68	6.36	5.42	4.89	4.56

df for the unexplained variation (e.g. residual df or within df)	df for the explained variation (e.g. regression df or between df): 1	2	3	4	5
16	**4.49**	**3.63**	**3.24**	**3.01**	**2.85**
	8.53	6.23	5.29	4.77	4.44
17	**4.45**	**3.59**	**3.20**	**2.96**	**2.81**
	8.40	6.11	5.18	4.67	4.34
18	**4.41**	**3.55**	**3.16**	**2.93**	**2.77**
	8.29	6.01	5.09	4.58	4.25
19	**4.38**	**3.52**	**3.13**	**2.90**	**2.74**
	8.18	5.93	5.01	4.50	4.17
20	**4.35**	**3.49**	**3.10**	**2.87**	**2.71**
	8.10	5.85	4.94	4.43	4.10
21	**4.32**	**3.47**	**3.07**	**2.84**	**2.68**
	8.02	5.78	4.87	4.37	4.04
22	**4.30**	**3.44**	**3.05**	**2.82**	**2.66**
	7.95	5.72	4.82	4.31	3.99
23	**4.28**	**3.42**	**3.03**	**2.80**	**2.64**
	7.88	5.66	4.76	4.26	3.94
24	**4.26**	**3.40**	**3.01**	**2.78**	**2.62**
	7.82	5.61	4.72	4.22	3.90
25	**4.24**	**3.39**	**2.99**	**2.76**	**2.60**
	7.77	5.57	4.68	4.18	3.85
30	**4.17**	**3.32**	**2.92**	**2.69**	**2.53**
	7.56	5.39	4.51	4.02	3.70
35	**4.12**	**3.27**	**2.87**	**2.64**	**2.49**
	7.42	5.27	4.40	3.91	3.59
40	**4.08**	**3.23**	**2.84**	**2.61**	**2.45**
	7.31	5.18	4.31	3.83	3.51
45	**4.06**	**3.20**	**2.81**	**2.58**	**2.42**
	7.23	5.11	4.25	3.77	3.45
50	**4.03**	**3.18**	**2.79**	**2.56**	**2.40**
	7.17	5.06	4.20	3.72	3.41

Table D.8 *Values of the chi-square distribution* (χ^2) *for use in frequency analysis for associations between variables, goodness of fit tests, Kruskal–Wallis one way analysis of variance using ranks, two-way analysis of variance using ranks, and Friedman's matched group analysis of variance using ranks. If this table is used for a test of association or independence (see Box 6.2), then the df are* $(r - 1)(c - 1)$. *If the table is used for a goodness of fit test (see Chapter 6), then the degrees of freedom are the number of categories minus* 1. *Where the table is being used as part of a Kruskal–Wallis one-way analysis of variance using ranks (see Box 7.4) or for Friedman's matched group analysis of variance using ranks (see Box 7.9), then the degrees of freedom are* $k - 1$. *If the table is used as part of a Kruskal–Wallis two-way analysis of variance using ranks (see Box 7.8), then the degrees of freedom are those associated with the variable* $(r - 1$ *or* $c - 1)$ *or the interaction between the variables* $([r - 1][c - 1])$. *The results are significant if the calculated value of* X^2, H *or* F_r *is higher than the table value*

df	χ^2		df	χ^2	
	P = 0.05	P = 0.01		**P = 0.05**	P = 0.01
1	**3.841**	6.635	16	**26.296**	32.000
2	**5.991**	9.210	17	**27.587**	33.409
3	**7.815**	11.345	18	**28.869**	34.805
4	**9.488**	13.277	19	**30.144**	36.191
5	**11.070**	15.086	20	**31.410**	37.566
6	**12.592**	16.812	21	**32.671**	38.932
7	**14.067**	18.475	22	**33.924**	40.289
8	**15.507**	20.090	23	**35.172**	41.638
9	**16.919**	21.666	24	**36.415**	42.980
10	**18.307**	23.209	25	**37.652**	44.314
11	**19.675**	24.725	26	**38.885**	45.642
12	**21.026**	26.217	27	**40.113**	46.963
13	**22.362**	27.688	28	**41.337**	48.278
14	**23.685**	29.141	29	**42.557**	49.588
15	**24.996**	30.578	30	**43.773**	50.892

Table D.9 *Values of* F_{max} *for testing equality of variances. Values are given for* **P = 0.05**. *The results are significant if the calculated value of* F_{max} *is greater than the table value*

Number of samples being compared (k)

		2	3	4	5	6	7	8	9	10
	2	39.0	87.5	142	202	266	333	403	475	550
	3	15.4	27.8	39.2	50.7	62.0	72.9	83.5	93.9	104
Degrees of	4	9.60	15.5	20.6	25.2	29.5	33.6	37.5	41.1	44.6
freedom of	5	7.15	10.8	13.7	16.3	18.7	20.8	22.9	24.7	26.5
the smallest	6	5.82	8.38	10.4	12.1	13.7	15.0	16.3	17.5	18.6
sample	7	4.99	6.94	8.44	9.70	10.8	11.8	12.7	13.5	14.3
	8	4.43	6.00	7.18	8.12	9.03	9.78	10.5	11.1	11.7
	9	4.03	5.34	6.31	7.11	7.80	8.41	8.95	9.45	9.91
	10	3.72	4.85	5.67	6.34	6.92	7.42	7.87	8.28	8.66
	15	2.86	3.54	4.01	4.37	4.68	4.95	5.19	5.40	5.59
	20	2.46	2.95	3.29	3.54	3.76	3.94	4.10	4.24	4.37
	30	2.07	2.40	2.61	2.78	2.91	3.02	3.12	3.21	3.29

Table D.10 *Values of q (at P = 0.05) for use in calculating the minimum significant difference (MSD) using the Tukey, Tukey–Kramer and Nemenyi tests. Values are selected using the numbers of samples being compared (k) and the appropriate df: for the Tukey test (see Box 7.2) and the Tukey–Kramer Test (see Box 7.3), these are the within df; for the Nemenyi test (see Box 7.5) use df = ∞*

		Number of samples being compared (k)							
		3	4	5	6	7	8	9	10
	1	26.98	32.82	37.08	40.41	43.12	45.40	47.36	49.07
	2	8.33	9.80	10.88	11.74	12.44	13.03	13.54	13.99
	3	5.91	6.83	7.50	8.04	8.48	8.85	9.18	9.46
	4	5.04	5.76	6.29	6.71	7.05	7.35	7.60	7.83
	5	4.60	5.22	5.67	6.03	6.33	6.58	6.80	7.00
	6	4.34	4.90	5.31	5.63	5.90	6.12	6.32	6.49
	7	4.17	4.68	5.06	5.36	5.61	5.82	6.00	6.16
Degrees of freedom for MS_{within}	8	4.04	4.53	4.89	5.17	5.40	5.60	5.77	5.92
	9	3.95	4.42	4.76	5.02	5.24	5.43	5.60	5.74
	10	3.88	4.33	4.65	4.91	5.12	5.31	5.46	5.60
	11	3.82	4.26	4.57	4.82	5.03	5.20	5.35	5.49
	12	3.77	4.20	4.51	4.75	4.95	5.12	5.27	5.40
	13	3.74	4.15	4.45	4.69	4.89	5.05	5.19	5.32
	14	3.70	4.11	4.41	4.64	4.83	4.99	5.13	5.25
	15	3.67	4.08	4.37	4.60	4.78	4.94	5.08	5.20
	16	3.65	4.05	4.33	4.56	4.74	4.90	5.03	5.15
	17	3.63	4.02	4.30	4.52	4.71	4.86	4.99	5.11
	18	3.61	4.00	4.28	4.50	4.67	4.82	4.96	5.07
	19	3.59	3.98	4.25	4.47	4.65	4.79	4.92	5.04
	20	3.58	3.96	4.23	4.45	4.62	4.77	4.90	5.01
	24	3.53	3.90	4.17	4.37	4.54	4.68	4.81	4.92
	30	3.49	3.85	4.10	4.30	4.46	4.60	4.72	4.82
	40	3.44	3.79	4.04	4.23	4.39	4.52	4.64	4.74
	60	3.40	3.74	3.98	4.16	4.31	4.44	4.55	4.65
	∞	3.31	3.63	3.86	4.03	4.17	4.29	4.39	4.47

Table D.11 *Values of* **Q** *(at* **P** *= 0.05) for calculating the minimum significant difference (MSD) using Dunn's method. Values are selected using the numbers of samples being compared (**k**) (see Box 7.6)*

Number of samples (k)							
3	4	5	6	7	8	9	10
2.394	2.639	2.807	2.936	3.038	3.124	3.197	3.261

Table D.12 *Values of* F_r *for Friedman's matched group analysis of variance using ranks. The upper table values (in bold) are for* **P** *= 0.05, while the lower values are for* **P** *= 0.01. The results are significant if the calculated value of* F_r *is higher than the table value*

		Number of samples (k)				
Number of matched groups (g)		3	4	5	6	>6
	3	**6.000**	**7.400**	**8.533** for g ≥ 3	**11.070**	For all
		–	9.000	10.13	15.088	values of
	4	**6.500**	**7.800**	**8.800**		g see χ^2
		8.000	9.600	11.20		tables
	5	**6.400**	**7.800**	**8.960**		with df of
		8.400	9.960	11.68		k minus 1
	6	**7.000**	**7.600** for g > 5	**9.49**		(Table
		9.000	10.20	13.28		D.8)
	7	**7.143**	**7.800**			
		8.857	10.54			
	8	**6.250**	**7.650**			
		9.000	10.50			
	9	**6.222** for g > 8	**7.815**			
		9.556	11.345			
	10	**6.200**				
		9.600				
	11	**6.545**				
		9.455				
	12	**6.500**				
		9.500				
	13	**6.615**				
		9.385				
	>13	for g >13 **5.991** 9.210				

Appendix E
Answers to exercises

The answers in this section reflect those which should be obtained by doing the exercises by hand. For those using computer programs, full probabilities are also given in brackets, together with, where relevant, appropriate correction factors (including correction for tied ranks in some nonparametric tests).

Chapter 1

1.1

Variable	Nominal	Ordinal	Interval/ratio
Ratio of a tree's breadth to its height			✔
Level of trampling on a footpath (low, medium or high)		✔	
Of the invertebrates caught in a trap, the percentage that are herbivorous (plant eating)			✔
Number of lorries passing a census point within one hour			✔
Footpath construction (concrete, gravel, none)	✔		

1.2

Variable	Continuous	Discrete
Amount of rainfall in a day (mm)	✔	
Number of people with an interest in environmental issues		✔
Percentage of leaves on a tree displaying fungal infection	✔	
The concentration of cadmium in roadside soils	✔	
Type of site (either reclaimed or not)		✔

1.3

Variable	Derived	Not derived
pH of soil	✔ *	
Number of households that own a car		✔
Percentage of tree canopy cover in a woodland	✔	
Estimate of tree canopy cover in a woodland (open, intermediate, closed)		✔
Type of soil (designated as clay, loam or sandy		✔

* pH is the logarithm of the reciprocal of the hydrogen ion concentration and so is effectively derived from the hydrogen ion concentration despite usually being measured directly.

1.4 (i) Stratified-random.

Advantage: incorporates the advantages of both of the methods below.

(ii) Random.

Advantage: reduces the opportunity for the researcher to unconsciously bias where the quadrats are placed.

Disadvantage: there may be areas of the meadow with different conditions that, by chance, are underrepresented.

(iii) Systematic.

Advantage: covers all of the meadow, taking into account the possibility that conditions across the site may differ.

Disadvantage: there may be an underlying pattern of plant diversity that has a 10 m periodicity.

1.5 He should take a single reading from each of 20 different large cars and from each of 20 different small cars.

Repeatedly sampling the same car is incorrect, since the data would be non-independent (pseudoreplication). Each of the cars may not be representative of the car population as a whole.

Chapter 2

2.1 (i)

x	f
0.70 – 0.74	1
0.75 – 0.79	3
0.80 – 0.84	3
0.85 – 0.89	4
0.90 – 0.94	3
0.95 – 0.99	4
1.00 – 1.04	1
1.05 – 1.09	1

(ii) Mass of hedgehogs in autumn ($n = 20$)

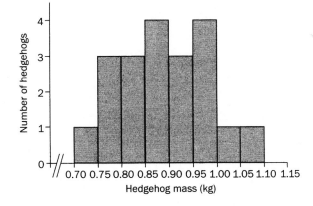

(iii)

	Spring	Autumn
Mean	0.667	0.889
Standard deviation	0.1090	0.0961
Standard error of the mean	0.0244	0.0215
Degrees of freedom	19	19
95% confidence limits of the mean	0.0510	0.0450

(iv) Mass of hedgehogs caught in autumn and spring. Bars are standard errors (*n* = 20 for each)   *or*   Mass of hedgehogs caught in autumn and spring. Bars are standard errors (*n* = 20 for each)

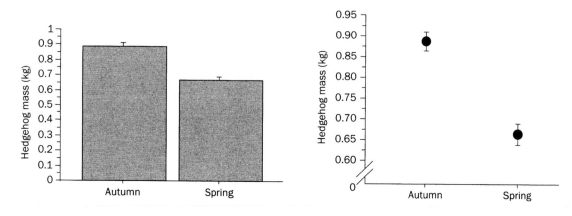

2.2 (i)

	Genetically modified food	*Intensive pesticide use*
Median	3	1
Lower quartile	2	1
Upper quartile	4	2

(ii) The level of public concern regarding the genetically modified food and intensive pesticide use (*n* = 20 in each case)

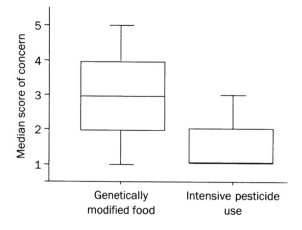

Chapter 3

3.1 (i) Percentage of leaves with fungus on each of 20 trees

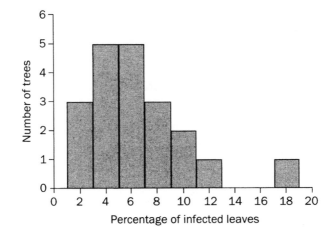

(ii) Percentage of leaves with fungus on each of 20 trees (arcsine-transformed data)

Degrees   *or*   Radians

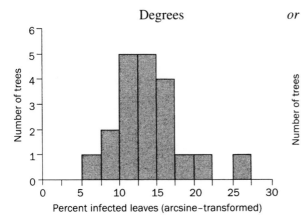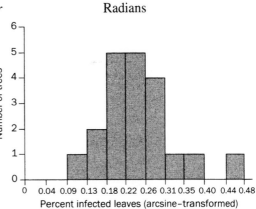

(iii)

	Degrees	*or*	*Radians*
Mean	13.78		0.24
Standard deviation	4.595		0.080
Standard error	1.028		0.018
Degrees of freedom	19		19
Lower 95% confidence level (L_{lower})	11.628		0.203
Upper 95% confidence level (L_{upper})	15.930		0.278

(iv)

Mean	5.6%
Lower 95% confidence level (L_{lower})	4.06%
Upper 95% confidence level (L_{upper})	7.53%

3.2 (i) There is no significant difference between the maximum temperatures in the urban and rural areas.

(ii) Reject the null hypothesis.

(iii) The urban area has a significantly higher maximum temperature than the rural site.

Chapter 4

4.1

	Rural	Urban
Mean suspended sediment	55.2 mg l^{-1}	75.5 mg l^{-1}
Standard deviation	17.58 mg l^{-1}	21.28 mg l^{-1}
$t = 2.326$	df = 18	$P < 0.05$ (= 0.0319)

There is significantly more suspended sediment in urban rivers than rural rivers.

4.2

	Active	Not in use
Median score	3	2
Sample size	15	15
Smaller U value = 60.5	$P < 0.05$ (= 0.0310 or 0.0256 if corrected for ties)	

Active badger sets are significantly more visible than those not in use.

4.3

The mean difference (d) = 8.0 µg g^{-1}		
$t = 2.419$	df = 5	$P > 0.05$ (= 0.0602)

There is no significant difference between the amount of copper in the humus and soil layers. Since all of the humus measurements were larger than the associated soil measurements, the lack of significant difference may reflect the small sample size involved.

4.4

$T = 34$	$n = 17$	$P < 0.05$ (= 0.0442 or 0.0253 if corrected for ties)

Businesses were significantly more enthusiastic about improving the environmental impact of their products after receiving the results of the customer questionnaire.

4.5 (i) Ideally a paired t test would be employed (a Wilcoxon test could be used, but is more conservative when data are likely to be normally distributed).

(ii) A Mann–Whitney U test would be appropriate for this unpaired comparison of ranked scores.

(iii) A Wilcoxon matched-pairs test would be suitable for these paired samples of ranked scores.

(iv) Ideally a t test would be employed (a Mann–Whitney U test could be used, but is more conservative when data are likely to be normally distributed).

Chapter 5

5.1 (i) Negative relationship between infant mortality rate and the number of physicians in African countries

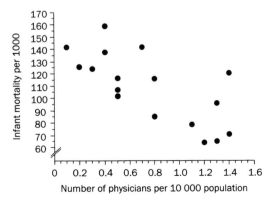

(ii)

Pearson's $r = -0.714$	df $= 15$	$P < 0.01 \, (= 0.0008)$

(iii) There is a highly significant negative relationship between the number of physicians and the infant mortality rate: the more physicians there are, the lower the infant mortality rate. Note that we are not assuming a cause and effect relationship.

5.2 (i)

$r_S = -0.482 \, (= -0.533$ if corrected for ties)	$n = 20$	$P < 0.05 \, (= 0.0203$ if corrected for ties)

(ii) There is a significant negative relationship between frequency of visit and amount the visitor is prepared to pay: the more frequent the visits, the less the visitor is prepared to pay.

5.3 (i)

$R^2 = 69.4\%$	Regression equation: Economy $= 22.39 - 5.862 \times$ Engine size (or $y = 22.39 - 5.862 \, x$)	
$F = 49.961$	df $= 1$ and 22	$P < 0.01 \, (< 0.0001)$

(ii) Negative relationship between engine size and fuel economy.

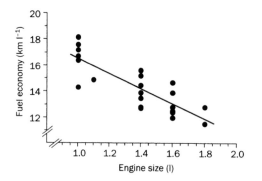

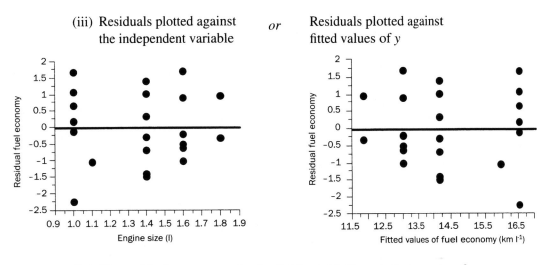

(iii) Residuals plotted against   *or*   Residuals plotted against
     the independent variable   fitted values of *y*

(iv) The residuals appear randomly distributed (although there are too few
     observations for the engine size of 1.8 litres to be completely sure of the
     distribution at that point). The distribution of the residuals therefore satisfies the
     assumptions of regression.

(v) There is a significant negative relationship between engine size and fuel
     economy: as engine size increases, fuel economy decreases.

Chapter 6

6.1 (i)

6.1 (i)	Pre-thicket	Thicket	High forest	Row totals
Present	7	10	5	22
Absent	5	1	8	14
				Grand total:
Column totals	12	11	13	36

(ii)	Pre-thicket	Thicket	High forest	Row totals
Present	7.333	6.722	7.944	22
Absent	4.667	4.278	5.056	14
				Grand total:
Column totals	12	11	13	36

(iii)

	Pre-thicket	Thicket	High forest
Present	0.0151	1.5983	1.0913
Absent	0.0238	2.5115	1.7149

(iv)

$X^2 = 6.9549$	df = 2	$P < 0.05$ (= 0.0309)

(v) There is a significant association between forest type and the presence or absence of long-tailed tits, with tits being more likely to be present in thickets (and absent from high forests).

6.2 (i)

	Observed	Expected	X^2
North	36	40	0.4
East	50	40	2.5
South	29	40	3.025
West	45	40	0.625

(ii)

total $X^2 = 6.55$	df = 3	$P > 0.05$ (= 0.088)

(iii) There is no significant departure from random orientation in lichens.

Chapter 7

7.1 (i) The highest variance is for minibuses (30.786).

The lowest variance is for cars (12.982).

$F_{max} = 2.37$, $k = 3$, df = 7, $P > 0.05$; therefore the variances are not significantly different.

(ii)	df	Sums of squares	Mean squares	F	P
Vehicle type	2	2629.750	1314.875	58.516	< 0.0001
Within	21	471.875	22.470		
Total	23	3101.625			

(iii)

Cars		Minibuses		Buses	
$\bar{x}$: 68.1 ppb	SE: 1.27 ppb	$\bar{x}$: 52.3 ppb	SE: 1.96 ppb	$\bar{x}$: 42.8 ppb	SE: 1.72 ppb

(iv) There is a significant difference between the NO_2 concentration in the different types of vehicle, with cars having the most and buses the least. Using Tukey's multiple comparisons:

$$MSD = q \times \sqrt{\frac{MS_{within}}{n}} = 3.5675 \times \sqrt{\frac{22.47}{8}} = 5.9789$$

The MSD is smaller than the difference between any pair of means, therefore all the means are significantly different from each other (Cars > Minibuses > Buses; $P < 0.05$)

7.2 (i)

Rank means			Kruskal–Wallis		
Up to 25	26 to 50	Over 51	df:	H:	P:
18.85	8.15	19.5	2	10.483 (corrected 10.984)	< 0.01 (exact, corrected 0.0041)

(ii) There is a significant difference among age groups as to enthusiasm towards recycling. Nemenyi's multiple comparisons,

$$MSD = q \times \sqrt{\frac{k(n_T - 1)}{12}} = 3.314 \times \sqrt{\frac{3 \times 31}{12}} = 9.226$$

reveal that those aged up to 25 and over 51 do not differ in their enthusiasm, but both age groups are significantly more enthusiastic than those aged 26–50 ($P < 0.05$): (up to 25, over 51) > 26 to 50.

7.3 (i)	df	Sums of squares	Mean squares	F	P
Region	1	0.011	0.011	5.136	< 0.05 (0.0377)
Location	1	0.013	0.013	5.799	< 0.05 (0.0285)
Interaction	1	0.00006845	0.00006845	0.031	> 0.05 (0.8632)
Within	16	0.036	0.002		
Total	19	0.060			

(ii) Effect of region and location on percentage loss of forest. Bars are standard errors.

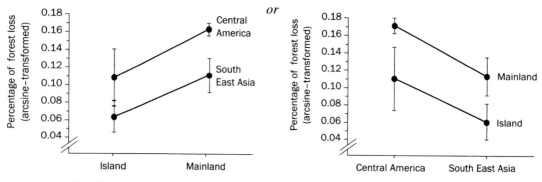

(iii) Central America has a significantly higher percentage forest loss than does South East Asia and mainland countries have significantly higher loss than islands. There is no significant interaction between region and location, thus the effects of region (Central America or South East Asia) are independent of those of location (island or mainland). Note that multiple comparisons are not needed when only two categories are being combined at a time.

7.4 (i)

	df	SS	MS	F	P
Route	1	883.600	883.600	14.286	0.0194
Region	4	6957.400	1739.350	28.122	0.0035
Remainder	4	247.000	61.850		
Total	9	8088.000			

(ii) River input is significantly higher than direct input ($\bar{x}_{River} = 42$; $\bar{x}_{Direct} = 23.2$), and there is a significant difference among regions. Tukey's multiple comparisons,

$$\text{MSD} = q \times \sqrt{\frac{MS_{remainder}}{n}} = 7.826 \times \sqrt{\frac{61.85}{2}} = 43.52$$

reveal that the North Sea East Coast has significantly higher nitrogen input than all the other regions ($P < 0.05$). None of the other regions differ from each other: North Sea East Coast > (Irish Sea, Celtic Sea, Atlantic, North Sea Channel)

7.5 (i)

	df	SS	MS	H	P
Party	1	99.225		0.7389	> 0.05
Timing	1	1416.1		10.545	< 0.01 (0.0016)
Interaction	1	133.25		0.992	> 0.05
Within	36	3588.95			
Total	39	5237.525	134.29555		

(ii) Rank means for number of positive words in response to questions regarding environmental issues from MPs from two political parties, before and after a general election

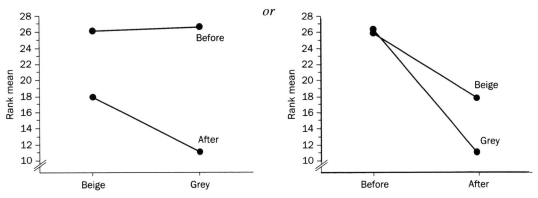

(iii) There is a significant effect of timing: politicians say fewer positive words about environmental issues after the general election. There is no significant difference between the parties in the number of positive words in politicians' responses, and there is no significant interaction between timing and political party. (Multiple comparisons are not necessary, because there is no interaction and only two categories in the significant factor.)

7.6 (i)

	Rank mean	Friedman's test			
Mammals	2.875	Number of samples (k)	Number of groups (g)	F_r:	P:
Birds	1.25	3	8	10.75	< 0.001
Reptiles/ Amphibians	1.875				

(ii) There is a significant difference in proportion of endangered species among animal types, with mammals having the greatest proportion of species at risk, followed by reptiles/amphibians and then birds. Nemenyi's multiple comparisons,

$$MSD = q \times \sqrt{\frac{k(k+1)}{12g}} = 3.314 \times \sqrt{\frac{3 \times 4}{12 \times 8}} = 1.172$$

reveal that there is a significantly higher proportion of mammals than birds at risk (Mammals > Birds at $P < 0.05$) but the proportion of reptiles/amphibians at risk is not significantly different from that of either mammals or birds.

Mammals   Reptiles/Amphibians   Birds

# References

Barnard, C., Gilbert F. and McGregor P. (1993) *Asking Questions in Biology*. Longman, Edinburgh.

Burt J. E. and Barber G. M. (1996) *Elementary Statistics for Geographers* (2nd edition). Guilford Press, London.

de Vaus D. A. (1996) *Surveys in Social Science* (4th edition), Social Research Today 5. UCL Press, London.

Gomez K. A. and Gomez A. A. (1984) *Statistical Procedures for Agricultural Research*. Wiley, New York.

Haining, R. (1993) *Spatial Data Analysis in the Social and Environmental Sciences*. Cambridge University Press, Cambridge.

Lindsay, J. M. (1997) *Techniques in Human Geography*. Routledge, London.

Manly B. F. J. (1994) *Multivariate Statistical Methods* (2nd edition). Chapman & Hall, London.

Mead R., Curnow R. N. and Hasted A. M. (1993) *Statistical Methods in Agriculture and Experimental Biology*. Chapman & Hall, London.

Meddis R. (1984) *Statistics Using Ranks: A Unified Approach*. Blackwell, Oxford.

Neave H. R. (1995) *Elementary Statistics Tables*. Routledge, London.

Nichols D. (1990) *Safety in Biological Fieldwork – Guidance Notes for Codes of Practice*. Institute of Biology, London.

Pigott J. and Watts S. (1996) Safety. In: S. Watts and L. Halliwell (eds) (1996) *Essential Environmental Science*. Routledge, London.

Robinson, G. M. (1998) *Methods and Techniques in Human Geography*. Wiley, Chichester.

Siegel S. and Castellan N. J. (1988) *Nonparametric Statistics for the Behavioural Sciences*. McGraw-Hill, London.

Sokal R. R. and Rohlf F. J. (1995) *Biometry* (3rd edition). W. H. Freeman & Co., New York.

Tabachnick B. G. and Fidell L. S. (1996) *Using Multivariate Statistics* (3rd edition). HarperCollins, New York.

Weber R. P. (1990) *Basic Content Analysis*. Sage Publications, Newbury Park, California.

Zar J. H. (1999) *Biostatistical Analysis* (4th edition). Prentice Hall, Upper Saddle River, NJ.

Index

Key to commonly used statistical tests Choosing a statistical test may sometimes seem rather bewildering. Until you have obtained experience in some of the methods, you will have to rely on advice and statistics textbooks. This table provides a key to many of the common tests. Start at the left-hand side of the table and work your way to the right. The starting point (first column) requires a decision about which of the three types of analysis questions (difference in central tendency, relationship between variables, or frequency analysis) you wish to ask. See Chapter 3 for an explanation of the broad categories of tests covered. After the first column there are pairs of questions which enable you to proceed (the first question in a pair begins with a capital letter and the second ends with a question mark). At each point you will need to decide on a direction based on some aspect of your data. Read the introduction at the relevant chapter to the test selected to be sure you have arrived at the correct test.

Category	Sub-question	Scale of measurement	Distribution	Transform?	Test
Differences in central tendency	2 samples — Data in matched pairs, or not?	Data measured on interval or ratio scales	Data normally distributed →	→	Paired t test
				Transform data →	Paired t test on transformed data
				or not? →	Wilcoxon matched-pairs test
		or on ordinal scale? →			Wilcoxon matched-pairs test
	(See Chapter 4) or not?	Data measured on interval or ratio scales	Data normally distributed →	→	t test
				Transform data →	t test on transformed data
				or not? →	Mann–Whitney U test
		or on ordinal scale? →			Mann–Whitney U test
	or more than 2 samples? — Data in matched groups, or not?	Data measured on interval or ratio scales	Data normally distributed →	→	Repeated measures or nested analysis of variance
				Transform data →	Repeated measures or nested analysis of variance on transformed data
				or not? →	Analysis of variance by ranks (Friedman's)
		or on ordinal scale? →			Analysis of variance by ranks (Friedman's)
	(See Chapter 7) or not?	Data measured on interval or ratio scales	Data normally distributed →	→	Analysis of variance (one-way, two-way, etc.)
				Transform data →	Analysis of variance on transformed data (one-way, two-way, etc.)
				or not? →	Analysis of variance by ranks (Kruskal–Wallis or two-way)
		or on ordinal scale? →			Analysis of variance by ranks (Kruskal–Wallis or two-way)
or relationships between variables (See Chapter 5)	Data measured on interval or ratio scales	Causal relationship suspected or predictive model required	Linear relationship between variables →	or not? →	Simple linear regression analysis
					Simple linear regression analysis on transformed data
		or not?	Linear relationship between variables →	→	Pearson's product moment correlation
				Transform data →	Pearson's product moment correlation on transformed data
				or not? →	Spearman's rank correlation
	or on ordinal scale?				Spearman's rank correlation
or frequency analysis (See Chapter 6)	Test observed frequencies against an expected distribution	Test against a model distribution →			Chi-square test for goodness of fit
		or for homogeneity? →			Chi-square test for homogeneity
	or test two observed distributions against each other? →				Chi-square for association/independence